基础　进阶　实训

MySQL
Database
Training Tutorial

MySQL
数据库项目化
实训教程

主　编◎许　雁　刘广耀
副主编◎王　堃　刘永波

同济大学 出版社
TONGJI UNIVERSITY PRESS
·上海·

内 容 提 要

MySQL 数据库是当前最为流行的开源数据库之一,它功能强大,已成为企业级数据库产品的首选。本书以网上购物系统的操作、管理为主线,巩固和深化数据库相关理论的学习和实践。本书共分为六个项目,内容涵盖 MySQL 数据库基础、操作数据库与数据表、查询系统数据、优化系统数据、数据库编程以及数据库安全性和可用性六个项目及下设的 18 个任务,每个项目中均配有若干典型实例。

本书可作为高等职业院校计算机相关专业及计算机教育培训机构专用教材,也可作为数据库开发爱好者的参考用书。

图书在版编目(CIP)数据

MySQL 数据库项目化实训教程/许雁,刘广耀主编. --上海:同济大学出版社,2022.12
ISBN 978-7-5765-0517-7

Ⅰ.①M… Ⅱ.①许… ②刘… Ⅲ.①SQL 语言-数据库管理系统-高等职业教育-教材 Ⅳ.①TP311.132.3

中国版本图书馆 CIP 数据核字(2022)第 231090 号

MySQL 数据库项目化实训教程

主 编 许 雁 刘广耀　　副主编 王 堃 刘永波
责任编辑　任学敏　　助理编辑　竺奕辰　　责任校对　徐逢乔　　封面设计　渲彩轩

出版发行	同济大学出版社　www.tongjipress.com.cn	
	(地址:上海市四平路 1239 号　邮编:200092　电话:021-65985622)	
经　　销	全国各地新华书店	
排　　版	南京月叶图文制作有限公司	
印　　刷	启东市人民印刷有限公司	
开　　本	787mm×1092mm　1/16	
印　　张	8.5	
字　　数	212 000	
版　　次	2022 年 12 月第 1 版	
印　　次	2022 年 12 月第 1 次印刷	
书　　号	ISBN 978-7-5765-0517-7	
定　　价	36.00 元	

本书若有印装质量问题,请向本社发行部调换　　版权所有　侵权必究

前　言

　　党的二十大报告指出："育人的根本在于立德。""统筹职业教育、高等教育、继续教育协同创新,推进职普融通、产教融合、科教融汇,优化职业教育类型定位。"本书严格遵循职业技能要求,结合专业人才培养方案,科学合理地优化教学内容体系,充分将项目化教学和技能训练融入教材中,体现教材的育人功能。

　　MySQL是目前比较流行的关系型数据库管理系统,由于其功能强大、运行速度快,且是开放源码,目前发布的版本较以往版本性能更优,因此深受用户欢迎。

　　本书采用项目驱动的方式,以网上购物系统为主线,通过MySQL数据库基础、操作数据库与数据表、查询系统数据、优化系统数据、数据库编程以及数据库的安全性和可用性六个项目及下设的18个任务构建数据库的知识和技能体系。为了巩固学习效果,在每个项目后都配备了课后习题和项目实训,使读者能够及时将所学知识应用到实际工作任务。

　　本书每个项目都以"知识储备—实战演练—课后实训"的模式设计,以目前最新的MySQL8.0版本为开发环境,在传授知识和提高技能的基础上进一步培养实践能力。

　　本书精心设计了网上购物系统的相关数据信息,结构紧凑,语言通俗易懂,注重理论联系实际,具有较强的实用性和可操作性。

本书由许雁、刘广耀任主编,王堃、刘永波任副主编,同时王雪娟、冯少艳、史红娟也参与了本书的编写。尽管编写团队在编写过程中尽了最大努力,但书中难免存在不足和疏漏之处,敬请读者提出宝贵意见和建议,我们将不胜感激。

编者

2022 年 9 月

目录

前言

项目一　MySQL 数据库基础 ········· 001

- 任务 1　认识数据库 ········· 002
  - 1.1.1　数据库的概念 ········· 002
  - 1.1.2　关系模型数据库 ········· 002
  - 1.1.3　结构化查询语言 SQL ········· 003
- 任务 2　安装与配置 MySQL 数据库 ········· 004
  - 1.2.1　MySQL 的安装与配置 ········· 004
  - 1.2.2　MySQL 的使用 ········· 008
  - 1.2.3　MySQL 的图形化管理工具 ········· 010
- 任务 3　设置 MySQL 字符集 ········· 012
  - 1.3.1　查看 MySQL 字符集 ········· 012
  - 1.3.2　修改字符集 ········· 014
- 课后习题 ········· 015
- 项目实训 ········· 015

项目二　操作数据库与数据表 ········· 017

- 任务 1　网上购物系统数据库设计 ········· 018
  - 2.1.1　系统功能分析 ········· 018
  - 2.1.2　建立实体—关系模型 ········· 020
- 任务 2　创建和操作数据库 ········· 021
  - 2.2.1　创建数据库 ········· 021
  - 2.2.2　查看和选择数据库 ········· 023
  - 2.2.3　修改数据库 ········· 024
  - 2.2.4　删除数据库 ········· 025
- 任务 3　MySQL 的存储引擎和数据类型 ········· 027
  - 2.3.1　MySQL 的存储引擎 ········· 027

  2.3.2 MySQL 的数据类型 ·· 027

 任务 4 创建和操作数据表 ··· 030
  2.4.1 创建数据表 ·· 030
  2.4.2 设置约束条件 ·· 034
  2.4.3 修改数据表 ·· 039
  2.4.4 删除数据表 ·· 044

 任务 5 插入、修改和删除系统数据 ·· 045
  2.5.1 插入数据 ··· 045
  2.5.2 修改数据 ··· 048
  2.5.3 删除数据 ··· 049

 课后习题 ·· 052
 项目实训 ·· 053

项目三 查询系统数据 ··· 054

 任务 1 查询单表数据 ·· 055
  3.1.1 SELECT 基本查询语句 ·· 055
  3.1.2 查询数据列 ·· 055
  3.1.3 查询数据行 ·· 057
  3.1.4 数据排序 ··· 062
  3.1.5 数据分组统计 ·· 063
  3.1.6 使用图形化管理工具实现单表查询 ··· 065

 任务 2 查询多表数据 ·· 068
  3.2.1 内连接查询 ·· 068
  3.2.2 外连接查询 ·· 068
  3.2.3 复合条件连接查询 ·· 070
  3.2.4 主查询与子查询 ··· 070

 课后习题 ·· 073
 项目实训 ·· 074

项目四 优化系统数据 ··· 075

 任务 1 索引 ··· 076
  4.1.1 索引概述 ··· 076
  4.1.2 创建和查看索引 ··· 077
  4.1.3 维护索引 ··· 080

 任务 2 视图 ··· 081

4.2.1　视图概述 ……………………………………………………… 081
　　4.2.2　创建和查看视图 ……………………………………………… 082
　　4.2.3　维护视图 ……………………………………………………… 085
　　4.2.4　通过视图操作数据 …………………………………………… 086
课后习题 ……………………………………………………………………… 088
项目实训 ……………………………………………………………………… 089

项目五　数据库编程 ……………………………………………………… 090

任务 1　存储过程 …………………………………………………………… 091
　　5.1.1　存储过程概述 ………………………………………………… 091
　　5.1.2　创建存储过程 ………………………………………………… 091
　　5.1.3　调用存储过程 ………………………………………………… 093
任务 2　存储函数 …………………………………………………………… 094
　　5.2.1　存储函数概述 ………………………………………………… 094
　　5.2.2　创建存储函数 ………………………………………………… 094
　　5.2.3　调用存储函数 ………………………………………………… 097
任务 3　触发器 ……………………………………………………………… 097
　　5.3.1　触发器概述 …………………………………………………… 097
　　5.3.2　创建触发器 …………………………………………………… 097
　　5.3.3　查看触发器 …………………………………………………… 099
　　5.3.4　删除触发器 …………………………………………………… 100
课后习题 ……………………………………………………………………… 100
项目实训 ……………………………………………………………………… 101

项目六　数据库系统的安全性和可用性 ………………………………… 102

任务 1　用户权限管理 ……………………………………………………… 103
　　6.1.1　用户权限 ……………………………………………………… 103
　　6.1.2　用户管理 ……………………………………………………… 104
　　6.1.3　权限管理 ……………………………………………………… 106
　　6.1.4　事务 …………………………………………………………… 107
任务 2　数据的备份与恢复 ………………………………………………… 108
　　6.2.1　数据备份 ……………………………………………………… 108
　　6.2.2　数据恢复 ……………………………………………………… 112
任务 3　使用日志备份和恢复数据 ………………………………………… 114
　　6.3.1　日志概述 ……………………………………………………… 115

6.3.2　错误日志 ……………………………………………………………… 115
6.3.3　二进制日志 …………………………………………………………… 116
6.3.4　通用查询日志 ………………………………………………………… 118
6.3.5　慢查询日志 …………………………………………………………… 120
课后习题 …………………………………………………………………………… 120
项目实训 …………………………………………………………………………… 121

参考答案 …………………………………………………………………………… 122

参考文献 …………………………………………………………………………… 126

项目一

MySQL 数据库基础

数据库技术是计算机科学技术的重要分支，广泛应用于事务处理、数据检索、人工智能等领域。MySQL 作为关系型数据库管理系统的典型代表，具有体积小、成本低、开放源码等优点，广泛应用于中小型网站中，是企业级数据库产品的首选。

本项目在介绍数据库的基本概念的基础上，通过在 Windows 平台上安装、配置 MySQL 数据库，使读者掌握 MySQL 数据库的使用方法。

学习目标

- 了解数据库的基本概念
- 了解结构化查询语言
- 掌握在 Windows 平台上安装 MySQL 数据库的方法
- 掌握 MySQL 的启动和登录
- 掌握 MySQL 字符集设置

素质目标

通过介绍当前国内外数据库现状，以及当前大数据时代背景下国内知名的计算机企业，引导学生树立科技强国的思想，凸显技术应用，鼓励科学创新，激发学生的爱国主义热情。数据库的安装与配置过程能培养学生严谨的工作作风、科学的思维方式，以及一丝不苟的工匠精神和职业素养。

任务1　认识数据库

任务描述

了解和掌握数据库的概念、数据库的类型和 SQL 语言的相关知识。

1.1.1　数据库的概念

1. 数据库

数据库(Database)是用来存放数据的仓库,具体就是指长期存储在计算机内、有组织、可共享的数据集合。

2. 数据库系统

数据库系统(Database System)由硬件、软件、数据库以及用户构成。数据库系统的硬件包括计算机的主机、键盘、显示器和外围设备等。数据库系统软件包括数据库管理系统、操作系统和各种高级语言处理程序等。数据库管理系统是数据库系统的核心,对数据库的一切操作都是在数据库管理系统中进行的。用户是管理、开发和使用数据库的主体,通常可以分为终端用户、应用程序开发人员和数据库管理员三种类型。

3. 数据库管理系统

数据库管理系统(Database Management System)是一种操作和管理数据库的软件,使用它可以创建、使用和维护数据库,对数据库进行统一的管理和控制,保证数据库的安全性和完整性。目前常见的关系型数据库管理系统有 MySQL、Oracle、DB2、SQL Server 等。

1.1.2　关系模型数据库

1. 关系模型

关系模型以二维表来表示实体与实体之间的联系,其操作的对象和结果都是二维表。关系模型数据库就是若干个二维表的集合,例如商品关系(表 1-1)。

表 1-1　商品关系

商品编号	名称	价格(元)	类目	库存量
1001	西游记	50.00	书籍	100 本
1002	苹果	5.50	水果	30 个
1003	连衣裙	200.00	服装	60 件

2. 关系型数据库的存储结构

关系型数据库采用二维表来实现数据存储,数据库中的表从逻辑结构上就是由若干行和列交叉形成的,其中表中的一行称为一个记录,描述一个具体事物的一组数据;表中的一列称为一个字段,表示表中对象的一个属性。字段是数据库中可以操作的最小单位。

例如学生成绩表(表 1-2)中的行描述一名学生的相关数据,列表示学生的一个属性。

表 1-2 学生成绩表

学号	姓名	性别	课程名称	成绩	学分
20210101	王伟	女	网页设计	83	4
20210303	李芳华	女	商务礼仪	88	2
20210431	张凯	男	软件工程	90	4

3. 常见的关系型数据库

常见的关系型数据库管理系统有 MySQL 数据库管理系统、Oracle 数据库管理系统、DB2 数据库管理系统和 SQL Server 数据库管理系统等。

(1) MySQL 数据库管理系统

MySQL 是开放源码的数据库管理系统,属于 Oracle 公司。MySQL 具有性能好、成本低、可靠性好、可跨平台等特点,目前广泛应用于互联网行业,例如,百度、网易、新浪等互联网公司都用到了 MySQL。

(2) Oracle 数据库管理系统

Oracle 数据库系统是关系型数据库,不仅具有完整的数据管理功能,还是一个分布式数据库系统,支持各种分布式功能。Oracle 具有可移植性好、使用方便、开发工具的界面友好等特点,功能齐全。

(3) DB2 数据库管理系统

DB2 是国际商业机器(IBM)公司出品的关系型数据库管理系统,具有较好的可伸缩性。从大型机到单用户环境均可支持,应用于常见的服务器操作系统平台。DB2 具有很好的网络支持功能,适用于大型分布式应用系统。

(4) SQL Server 数据库管理系统

SQL Server 是微软(Microsoft)公司推出的关系型数据库管理系统,广泛应用于电子商务、银行等行业,具有可伸缩性、可用性、可靠性等特点,使系统管理和数据库管理更加直观、简单。

1.1.3 结构化查询语言 SQL

结构化查询语言(Structured Query Language,SQL)是目前被广泛使用的关系型数据库的标准语言,是非过程化编程语言。SQL 由于简单易学,且功能丰富,已被国际标准化组织确定为关系型数据库语言的国际标准。SQL 语言根据功能的不同被划分为数据定义语言、数据操作语言和数据控制语言。

(1) 数据定义语言

数据定义语言(Data Definition Language,DDL)用于创建数据库和数据库对象,为数据库操作提供对象。常用的语句关键字包括 CREATE、ALTER、DROP。CREATE 用于创建数据库对象,ALTER 用于修改数据库对象,DROP 用于删除数据库对象。

(2) 数据操作语言

数据操作语言(Data Manipulation Language,DML)用于操作数据库中的数据。常用的语句关键字包括 INSERT、UPDATE、DELETE、SELECT。INSERT 用于插入数据,UPDATE 用于修改数据,DELETE 用于删除数据,SELECT 用于查询数据。

(3) 数据控制语言

数据控制语言(Data Control Language,DCL)用于实现对象的访问权限和对数据库操作事务的控制。常用的语句关键字包括 GRANT、REVOKE、COMMIT、ROLLBACK。GRANT 用于给用户授予权限,REVOKE 用于收回用户权限,COMMIT 用于提交事务,ROLLBACK 用于撤销事务,即回滚到事务开始之前的状态。

任务 2　安装与配置 MySQL 数据库

任务描述

了解 MySQL 的安装与配置流程,并会使用命令和图形化管理工具操作数据库。

1.2.1　MySQL 的安装与配置

MySQL 支持 Windows、Linux、MacOS 等多种操作系统,本任务介绍在 Windows 10 操作系统中进行 MySQL 安装与配置的流程。

MySQL 的安装步骤如下。

(1) 下载 MySQL。登录官网下载安装文件,本书选择的安装版本为 8.0.30。打开浏览器,在地址栏输入"MySQL∷Download MySQL Community Server",按回车键进入下载页面,如图 1-1 所示。

单击右侧"Download"按钮,跳转到下一页面,如图 1-2 所示。

(2) 单击"No thanks,just start my download",直接下载并安装文件。也可以单击"Login"按钮进行 Oracle 账号申请,下载安装包进行安装。

单击"mysql-installer-community-8.0.30.0.msi"进行安装;安装流程如图 1-3 至图 1-5 所示。

项目一　MySQL 数据库基础

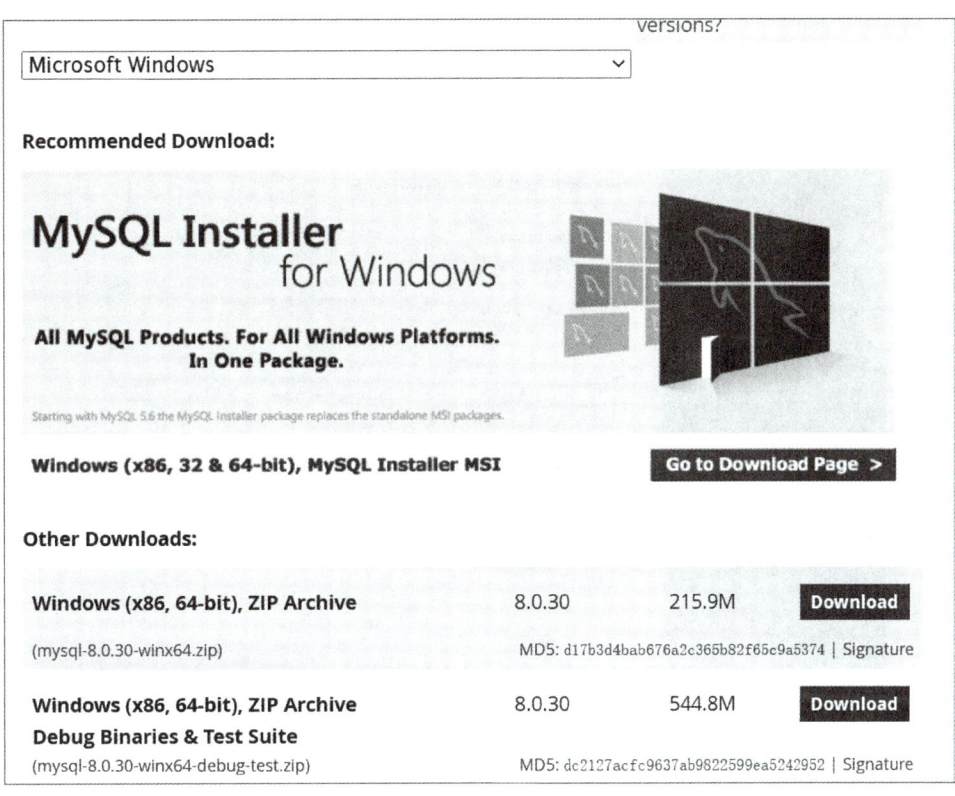

图 1-1　下载页面 1

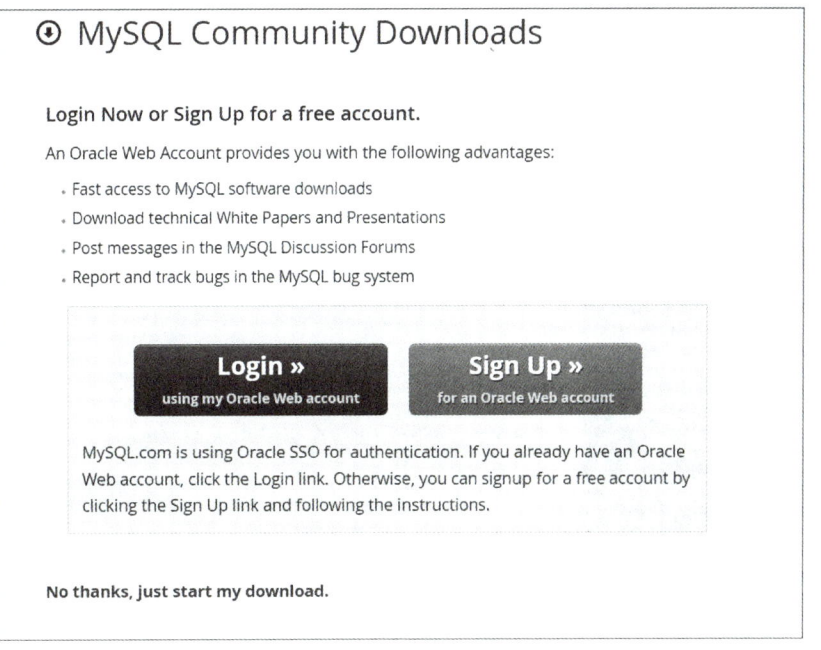

图 1-2　下载页面 2

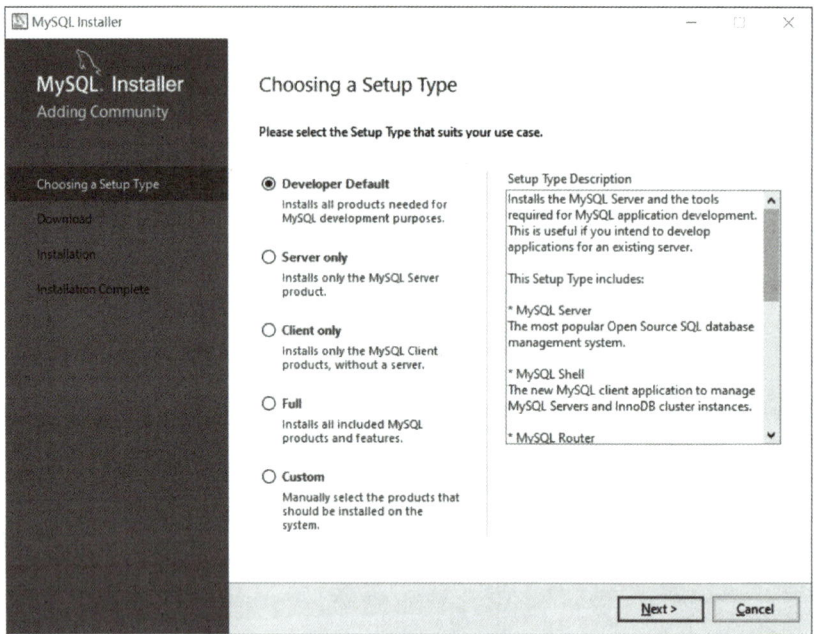

图 1-3　安装流程 1

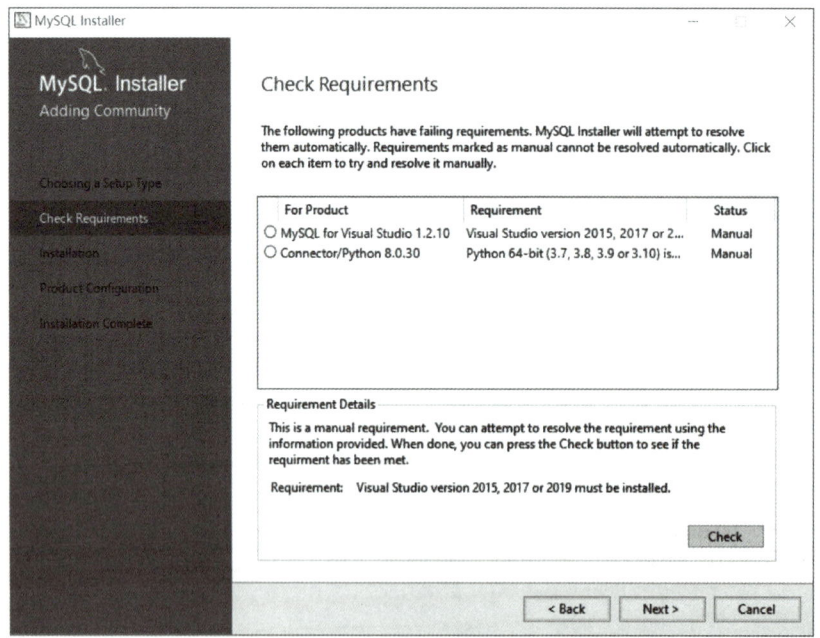

图 1-4　安装流程 2

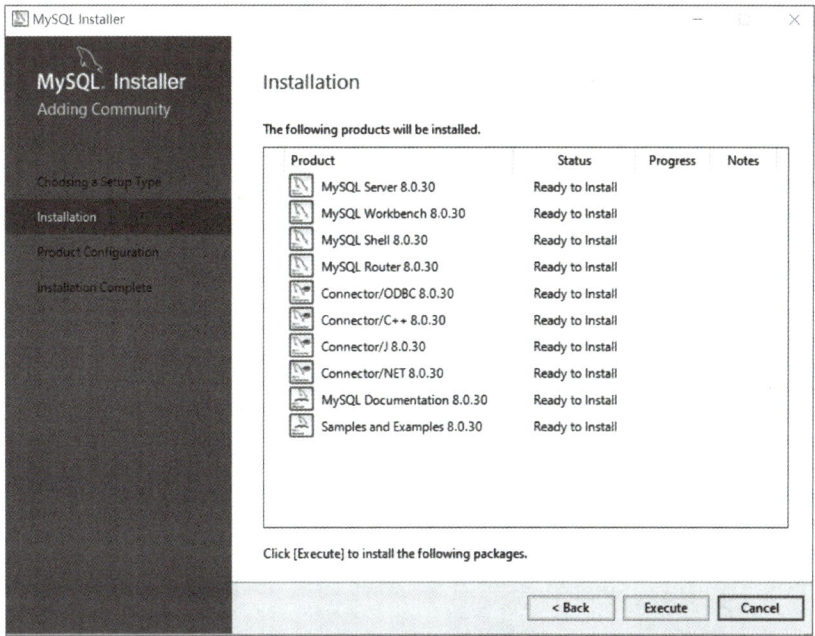

图 1-5　安装流程 3

（3）安装完成后进入安装完成界面（图 1-6），在此流程中，需要设置密码。

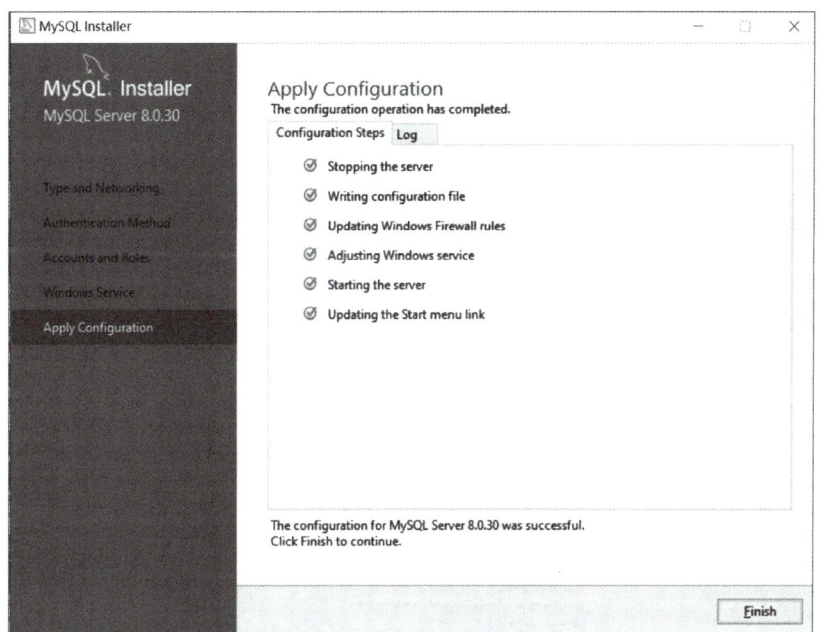

图 1-6　安装完成界面

1.2.2　MySQL 的使用

MySQL 安装完成后,需要启动服务,客户端才能登录 MySQL 服务器。

1. 启动和停止 MySQL 服务

用户要启动 MySQL 服务,共有两种方法,一种可以通过系统命令行来完成,另一种可以通过 Windows 服务器来完成。由于在安装时,已经将 MySQL 安装为 Windows 服务,所以可以通过 Windows 服务器直接启动 MySQL 服务。

(1) 通过 Windows 服务管理器启动和停止 MySQL 服务

使用 Windows 服务管理器可以启动和停止 MySQL 服务,打开"计算机管理"→"服务和应用程序"→"服务",找到 MySQL,单击右键或双击以启动服务或停止服务,如图 1-7、图 1-8 所示。

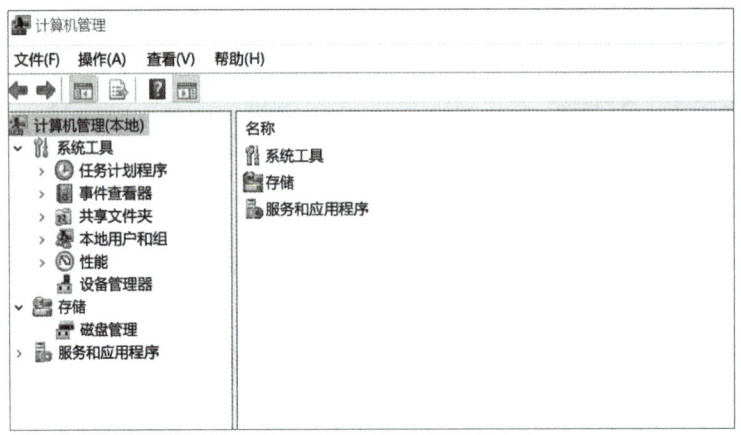

图 1-7　打开计算机管理界面

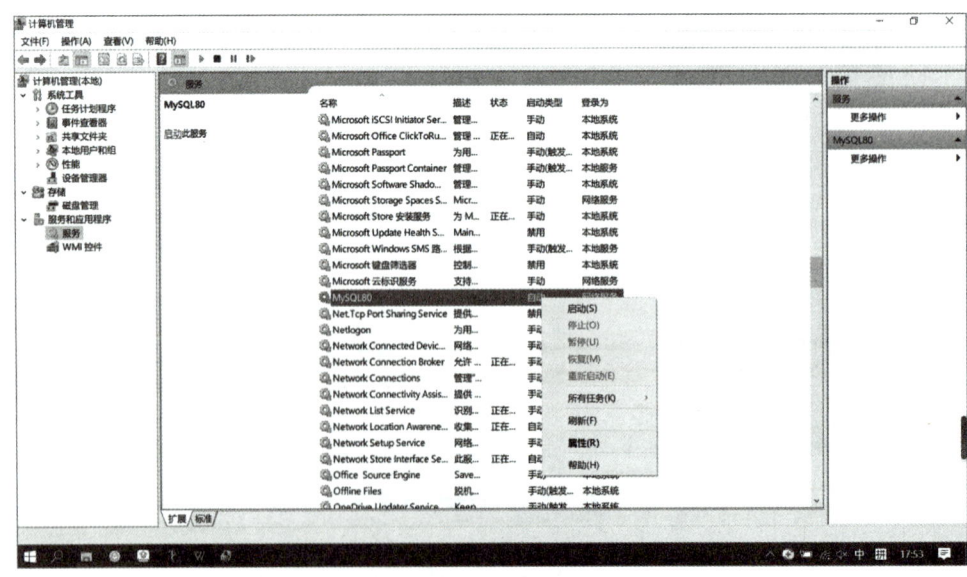

图 1-8　启动或停止服务

（2）通过操作系统命令启动和停止 MySQL 服务

单击"开始"→单击"运行"→输入"cmd"命令→回车。打开命令提示符窗口，输入启动命令"net start mysql80"，启动 MySQL，如图 1-9 所示。

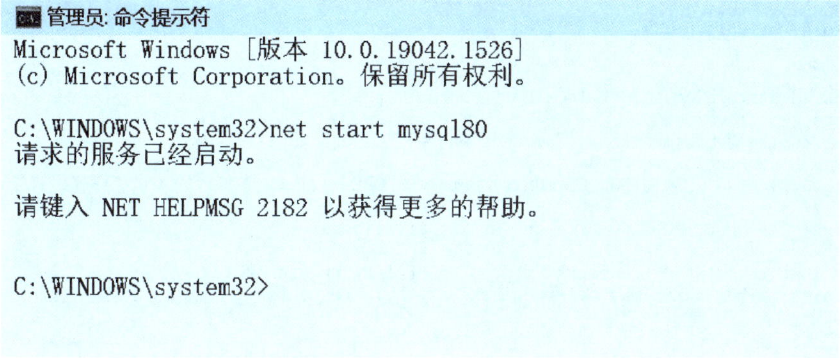

图 1-9　启动 MySQL 服务

输入停止 MySQL 命令"net stop mysql80"，停止 MySQL，如图 1-10 所示。

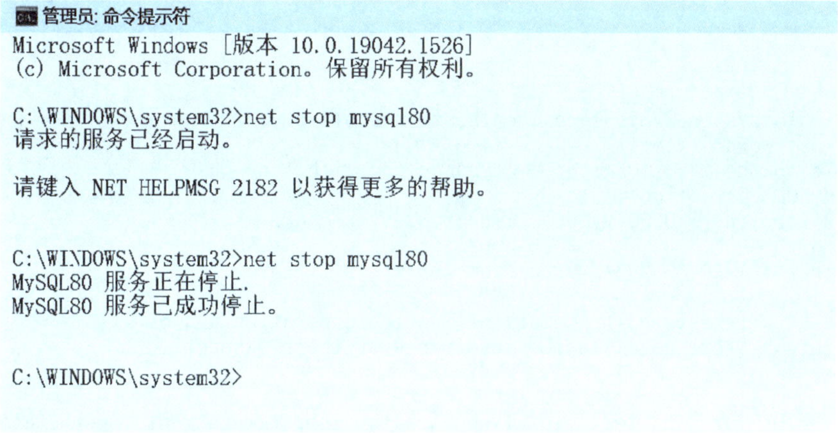

图 1-10　停止 MySQL 服务

2. 登录和退出 MySQL 服务器

MySQL 服务器启动后，可以通过客户端登录服务器，通过命令就可以操作和管理数据库及对象。在命令窗口输入的登录命令格式为：

mysql -h hostname -u username -p

说明如下：

"-h"指后面的参数为服务器主机地址，当客户端与服务端在同一台计算机上时，可以使用"localhost"，如果是本机登录，就可以省略该参数。"-u"表示后面的参数为登录 MySQL 服务的用户名，此处为"root"。"-p"表示后面的参数为指定用户的登录密码，但密码不需要在本行输入。按回车键确认后，系统会提示"Enter password"，输入设定密码，验证通过即可登录。登录成功后会加载 MySQL 服务器的欢迎和说明信息，并转入命令提示符"mysql>"，如图 1-11 所示。

```
C:\WINDOWS\system32>net start mysql80
MySQL80 服务正在启动.
MySQL80 服务已经启动成功。

C:\WINDOWS\system32>mysql -h localhost -u root -p
Enter password: *******
Welcome to the MySQL monitor.  Commands end with ; or \g.
Your MySQL connection id is 8
Server version: 8.0.30 MySQL Community Server - GPL

Copyright (c) 2000, 2022, Oracle and/or its affiliates.

Oracle is a registered trademark of Oracle Corporation and/or its
affiliates. Other names may be trademarks of their respective
owners.

Type 'help;' or '\h' for help. Type '\c' to clear the current input statement.
```

图 1-11　MySQL 服务器登录成功

退出 MySQL 需要在命令行窗口输入以下任意一个命令："exit""quit"或"\q"，如图 1-12 所示。

```
C:\WINDOWS\system32>mysql -h localhost -u root -p
Enter password: *******
Welcome to the MySQL monitor.  Commands end with ; or \g.
Your MySQL connection id is 8
Server version: 8.0.30 MySQL Community Server - GPL

Copyright (c) 2000, 2022, Oracle and/or its affiliates.

Oracle is a registered trademark of Oracle Corporation and/or its
affiliates. Other names may be trademarks of their respective
owners.

Type 'help;' or '\h' for help. Type '\c' to clear the current input statement.

mysql> exit;
Bye
```

图 1-12　退出 MySQL 服务器

1.2.3　MySQL 的图形化管理工具

对于初学 MySQL 的用户而言，使用命令窗口操作有一定困难，因此，许多公司开发了图形化管理工具，对数据库进行可视化操作，极大地提高了数据库操作和管理的效率。

MySQL 图形化管理工具很多，目前应用较广泛的有 Navicat for MySQL、MySQL WorkBench 和 SQLyog 等。由于各种图形化工具在数据库管理中有一定相似性，这里主要介绍 Navicat for MySQL。它有很好的用户界面，不仅可以用于访问、配置、控制和管理服务器中的所有对象和组件,而且可以将图形化工具和脚本编辑器融合,为用户提供数据库的管理、维护、查询等功能。

1. Navicat 的下载与安装

在浏览器地址栏中输入网址"http://www.navicat.com.cn",进入下载页面,选择"Navicat for MySQL",如图 1-13 所示。

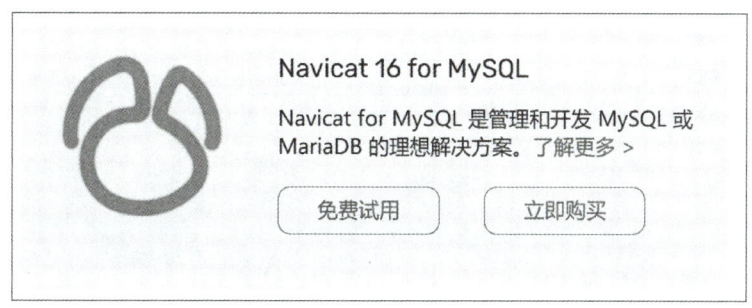

图 1-13　Navicat 的下载

下载安装程序,并选择安装,直至弹出"完成"按钮,如图 1-14 所示。

图 1-14　Navicat 的安装

2. 使用 Navicat 连接 MySQL

Navicat 是一个客户端软件,若要操作 MySQL,需要与之建立连接。步骤如下:
打开 Navicat 客户端软件,选择"连接"命令,弹出窗口如图 1-15 所示。

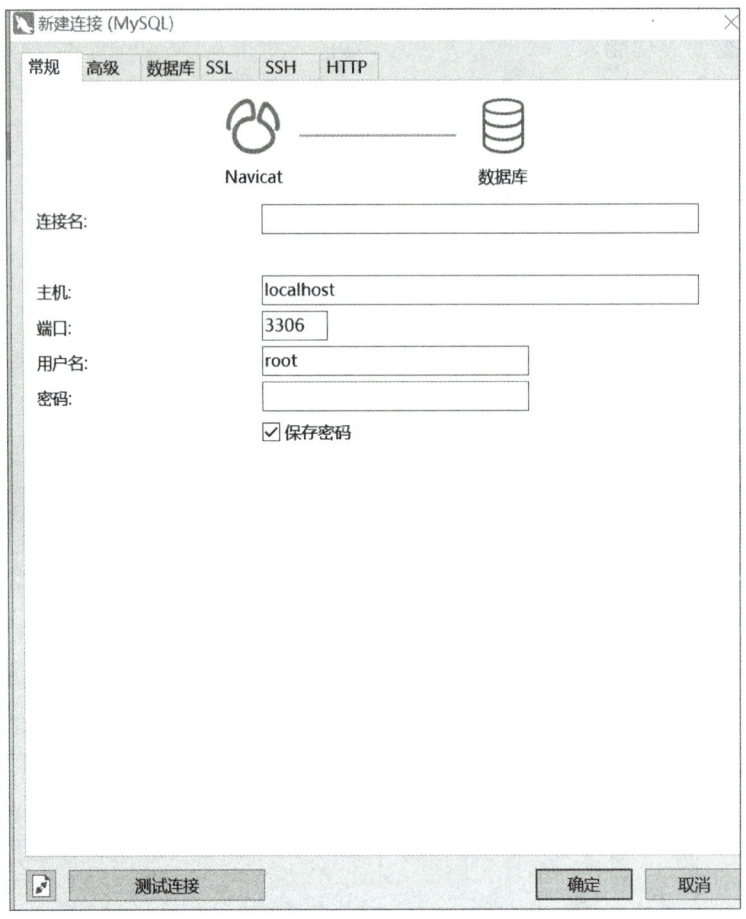

图 1-15　连接 MySQL

输入正确的连接名、主机名、端口、用户名和密码后,单击"确定"按钮就可以连接了。若本机也可以直接输入密码,单击"确定"按钮也可以连接成功。

任务 3　设置 MySQL 字符集

🔹 任务描述

了解 MySQL 支持的字符集,掌握如何设置和修改字符集。

1.3.1　查看 MySQL 字符集

1. 查看 MySQL 常用字符集

MySQL 支持多种字符集,其中常用的字符集包括 Latin1、UTF-8 和 gbk 等,其中

MySQL8.0 默认的字符集是 UTF-8mb4,执行相应语句,可以查看 MySQL 中可用的字符集。

输入以下命令语句:

SHOW CHARACTER SET；

结果如图 1-16 所示。

```
| ascii     | US ASCII                       | ascii_general_ci      | 1 |
| big5      | Big5 Traditional Chinese       | big5_chinese_ci       | 2 |
| binary    | Binary pseudo charset          | binary                | 1 |
| cp1250    | Windows Central European       | cp1250_general_ci     | 1 |
| cp1251    | Windows Cyrillic               | cp1251_general_ci     | 1 |
| cp1256    | Windows Arabic                 | cp1256_general_ci     | 1 |
| cp1257    | Windows Baltic                 | cp1257_general_ci     | 1 |
| cp850     | DOS West European              | cp850_general_ci      | 1 |
| cp852     | DOS Central European           | cp852_general_ci      | 1 |
| cp866     | DOS Russian                    | cp866_general_ci      | 1 |
| cp932     | SJIS for Windows Japanese      | cp932_japanese_ci     | 2 |
| dec8      | DEC West European              | dec8_swedish_ci       | 1 |
| eucjpms   | UJIS for Windows Japanese      | eucjpms_japanese_ci   | 3 |
| euckr     | EUC-KR Korean                  | euckr_korean_ci       | 2 |
| gb18030   | China National Standard GB18030| gb18030_chinese_ci    | 4 |
| gb2312    | GB2312 Simplified Chinese      | gb2312_chinese_ci     | 2 |
| gbk       | GBK Simplified Chinese         | gbk_chinese_ci        | 2 |
| geostd8   | GEOSTD8 Georgian               | geostd8_general_ci    | 1 |
| greek     | ISO 8859-7 Greek               | greek_general_ci      | 1 |
| hebrew    | ISO 8859-8 Hebrew              | hebrew_general_ci     | 1 |
| hp8       | HP West European               | hp8_english_ci        | 1 |
| keybcs2   | DOS Kamenicky Czech-Slovak     | keybcs2_general_ci    | 1 |
| koi8r     | KOI8-R Relcom Russian          | koi8r_general_ci      | 1 |
| koi8u     | KOI8-U Ukrainian               | koi8u_general_ci      | 1 |
| latin1    | cp1252 West European           | latin1_swedish_ci     | 1 |
| latin2    | ISO 8859-2 Central European    | latin2_general_ci     | 1 |
| latin5    | ISO 8859-9 Turkish             | latin5_turkish_ci     | 1 |
| latin7    | ISO 8859-13 Baltic             | latin7_general_ci     | 1 |
| macce     | Mac Central European           | macce_general_ci      | 1 |
| macroman  | Mac West European              | macroman_general_ci   | 1 |
| sjis      | Shift-JIS Japanese             | sjis_japanese_ci      | 2 |
| swe7      | 7bit Swedish                   | swe7_swedish_ci       | 1 |
| tis620    | TIS620 Thai                    | tis620_thai_ci        | 1 |
| ucs2      | UCS-2 Unicode                  | ucs2_general_ci       | 2 |
| ujis      | EUC-JP Japanese                | ujis_japanese_ci      | 3 |
| utf16     | UTF-16 Unicode                 | utf16_general_ci      | 4 |
| utf16le   | UTF-16LE Unicode               | utf16le_general_ci    | 4 |
| utf32     | UTF-32 Unicode                 | utf32_general_ci      | 4 |
| utf8mb3   | UTF-8 Unicode                  | utf8mb3_general_ci    | 3 |
| utf8mb4   | UTF-8 Unicode                  | utf8mb4_0900_ai_ci    | 4 |
```

图 1-16　输入命令语句

2. 查看不同级别字符集

MySQL 支持服务器(server)、数据库(database)、数据表(data table)和字段(field)四个层次的字符集设置,若要查看当前服务器级别的设置方法,可在命令行中输入以下语句(图 1-17):

SHOW VARIABLES LIKE 'character_set_server';

```
mysql> SHOW VARIABLES LIKE 'character_set_server';
+----------------------+--------+
| Variable_name        | Value  |
+----------------------+--------+
| character_set_server | utf8mb4|
+----------------------+--------+
1 row in set, 1 warning (0.03 sec)
```

图 1-17　查看服务器级别的设置方法

若要查看数据库级别的设置方法,可在命令行中输入以下语句(图 1-18):
SHOW VARIABLES LIKE 'character_set_database';

```
mysql> SHOW VARIABLES LIKE 'character_set_database';
+------------------------+--------+
| Variable_name          | Value  |
+------------------------+--------+
| character_set_database | utf8mb4|
+------------------------+--------+
1 row in set, 1 warning (0.00 sec)
```

图 1-18　查看数据库级别的设置方法

1.3.2　修改字符集

通过修改配置文件(my.ini)或设置系统变量可以实现字符集的设置和管理。系统启动时,默认字符集为 UTF-8mb4,该字符集是国际编码,能够实现对英文和中文的编码,因此,建议不要修改服务器端的字符集。但客户端的字符集取决于客户需求,所以本小节介绍如何修改客户端字符集。

若要设置或修改数据库中的变量,可在命令行中输入以下语句(图 1-19):
SET　变量名=值;

```
mysql> SET character_set_client=utf8mb4;
Query OK, 0 rows affected (0.00 sec)

mysql> SET character_set_connection=utf8mb4;
Query OK, 0 rows affected (0.00 sec)

mysql> SHOW VARIABLES LIKE 'CHAR%';
+--------------------------+-------------------------------------------------+
| Variable_name            | Value                                           |
+--------------------------+-------------------------------------------------+
| character_set_client     | utf8mb4                                         |
| character_set_connection | utf8mb4                                         |
| character_set_database   | utf8mb4                                         |
| character_set_filesystem | binary                                          |
| character_set_results    | gbk                                             |
| character_set_server     | utf8mb4                                         |
| character_set_system     | utf8mb3                                         |
| character_sets_dir       | C:\Program Files\MySQL\MySQL Server 8.0\share\charsets\|
+--------------------------+-------------------------------------------------+
8 rows in set, 1 warning (0.01 sec)
```

图 1-19　设置或修改数据库中的变量

课后习题

一、选择题

1. 数据库系统的核心是（　　）。
   A. 数据库　　　　　　　　　　　B. 数据
   C. 数据库管理系统　　　　　　　D. 数据库管理员
2. SQL 语言具有（　　）功能。
   A. 数据定义、数据操作、数据管理　　　B. 数据定义、数据控制、数据操作
   C. 数据规范、数据操作、数据管理　　　D. 数据定义、数据规范、数据操作
3. MySQL 是（　　）数据库管理系统。
   A. 层次型　　　B. 关系型　　　C. 网状型　　　D. 树型
4. 在数据库中存储的是（　　）。
   A. 信息　　　　　　　　　　　　B. 数据库
   C. 数据以及数据之间的联系　　　D. 数据库管理员
5. 负责数据库查询操作的数据库语言是（　　）。
   A. 数据定义语言　　　　　　　　B. 数据控制语言
   C. 数据管理语言　　　　　　　　D. 数据操作语言
6. MySQL 系统的默认配置文件是（　　）。
   A. my.ini　　　　　　　　　　　B. my-larger.ini
   C. my-huge.ini　　　　　　　　 D. my-small.ini

二、简答题

1. 简述常用的关系型数据库管理系统以及它们各自的特点。
2. 简述数据库、数据库系统和数据库管理系统之间的关系。

项目实训

1. 实训任务
（1）安装、配置和试用 MySQL。
（2）安装和使用 Navicat。
（3）MySQL 字符集的查看与修改。

2. 实训目的
（1）能正确安装和配置 MySQL 数据库。
（2）能使用操作系统命令和 Windows 服务启动和停止 MySQL 服务。
（3）能使用 Navicat 图形化管理工具操作 MySQL。

（4）能正确设置 MySQL 字符集。

3. 实训内容

（1）通过 MySQL 官网，下载并安装 MySQL。

（2）使用命令行工具启动和停止 MySQL 服务。

（3）通过 Navicat 官网，下载并安装 Navicat。

（4）分别使用命令行和 Navicat 图形化管理工具登录和退出 MySQL。

（5）使用 SQL 语句查看 MySQL 服务器的默认状态信息。

（6）使用 SQL 语句修改 MySQL 客户端字符集。

项目二

操作数据库与数据表

　　数据库是存储数据的仓库,数据库系统的设计是为了构造最优的数据模式,有效存储数据,满足用户的应用需求。数据表是数据库的操作对象。创建数据表时,需要选择存储引擎和字段的数据类型。软件开发人员应具备数据库与数据表的操作能力,例如数据的创建、修改、删除等。

　　本项目以网上购物系统为例,介绍数据库的设计过程以及在 MySQL 中创建和维护数据库及数据表的基本操作。

学习目标

- 理解网上购物系统数据库的设计过程,掌握 E-R 模型图
- 掌握创建和维护数据库的操作
- 理解 MySQL 的存储引擎和数据类型
- 掌握创建和维护数据表的操作
- 能利用 SQL 语句和 Navicat 图形化管理工具进行数据的插入、更新、删除操作

素质目标

　　建立 E-R 模型既要有全局意识,又要注重细节,从整体到局部,有利于培养学生的科学思维和严谨的科学素养。熟练进行数据库和数据表基本操作需要注重细节,这能培养学生爱岗敬业的大国工匠精神,使学生从实践操作中养成严谨务实的科学态度以及细致认真的工作习惯。

任务1　网上购物系统数据库设计

📖 任务描述

本任务通过对网上购物系统进行数据库设计，让学生掌握数据库功能设计、流程设计及 E-R 模型的建立，使消费者能通过网上购物系统完成商品的选购、支付、评价等。

2.1.1 系统功能分析

网上购物系统通常包括管理员、用户、商品和订单四类实体，分为用户购物和信息管理两种功能，主要涉及管理员、用户。

用户购物的主要功能见表 2-1。

表 2-1　用户购物的主要功能

功能	内容	角色
商品展示	展示商品名称、类别、编号、价格等信息	管理员
商品购买	用户将商品添加至购物车，确认购买后生成订单	用户
用户管理	用户可以注册、登录、查看、维护、注销个人信息	用户

信息管理的主要功能见表 2-2。

表 2-2　信息管理的主要功能

功能	内容	角色
商品信息维护	商品信息的添加、修改和删除	管理员
用户信息维护	用户信息的添加、维护、查询	管理员
订单信息维护	订单信息的查询、撤销、数据统计等	管理员

网上购物系统流程如图 2-1 所示。

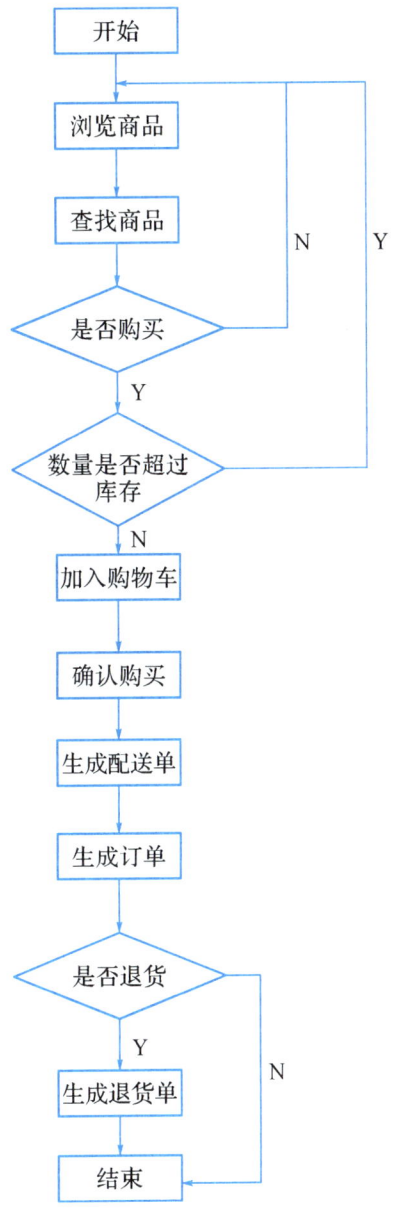

图 2-1 网上购物系统流程

2.1.2 建立实体—关系模型

（1）网上购物系统包括四类实体：管理员、用户、商品和订单，如图 2-2 所示。

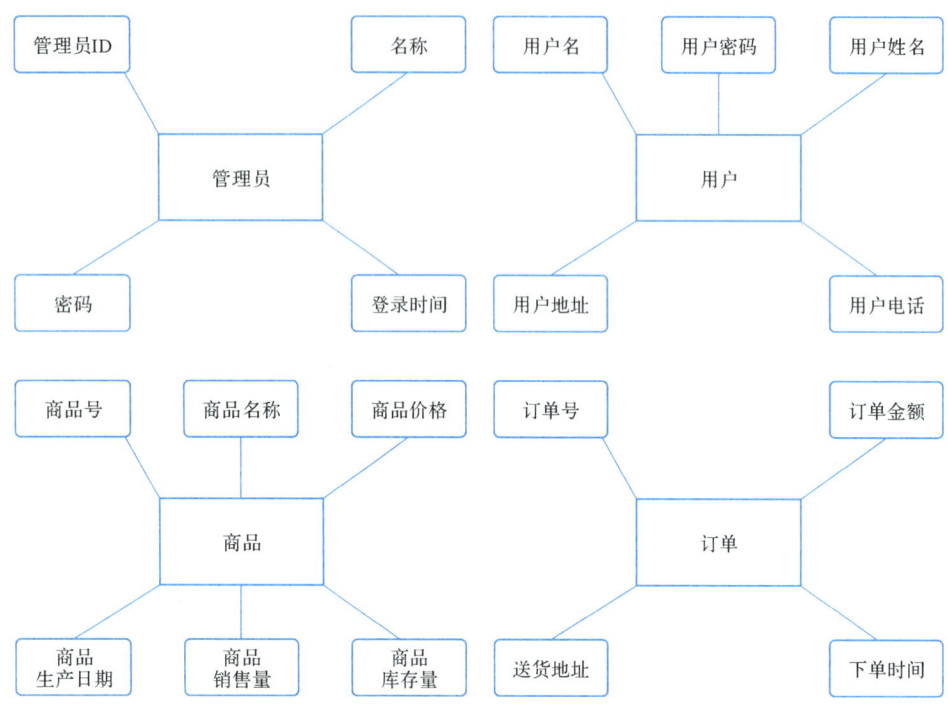

图 2-2　四类实体

（2）网上购物系统基本表组成见表 2-3、表 2-4。

表 2-3　客户信息表(示例)

用户名	密码	性别	电话	出生年月	电子邮箱	QQ 号码	用户积分
Jack123	＊＊＊＊	男	137＊＊＊＊＊＊＊	1987.11	＊＊@163.com	321093＊＊	100
Tom	＊＊＊＊	男	159＊＊＊＊＊＊＊	2001.05	＊＊@qq.com	445564＊＊	90
1388934	＊＊＊＊	女	187＊＊＊＊＊＊＊	1999.06	＊＊@126.com	134667＊＊	180

表 2-4　商品信息表(示例)

商品编号	商品类别	商品名称	商品价格(元)	商品库存	上架时间
001	食品	牛奶糖	1.2	2 000 颗	2022-06
002	服饰	西服	800.0	150 件	2022-01
003	书籍	《西游记》	50.0	800 本	2021-11

（3）模型是对现实世界的抽象，反映客观事物之间的关系。根据用户视角，数据模型从用户需求到实现可以按照以下关系进行：

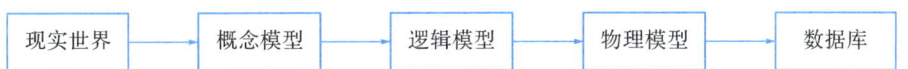

概念数据模型通常用实体—关系模型(Entity-Relationship model,E-R)来表示系统中的实体与关系,因此,可对网上购物系统进行 E-R 模型设计。网上购物系统实体—关系模型如图 2-3 所示。

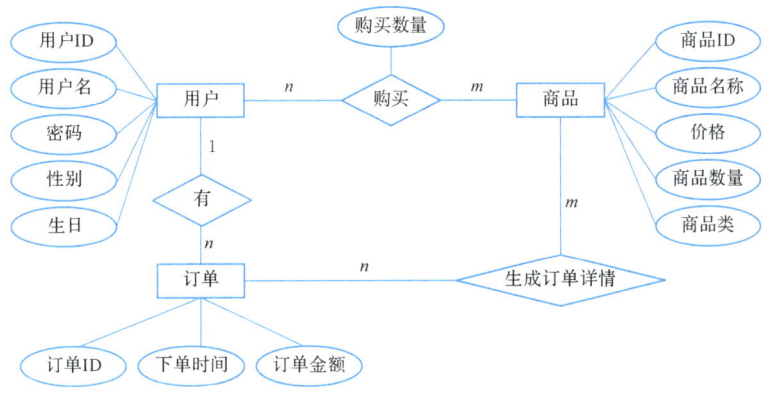

图 2-3 实体—关系模型

注：实体指客观存在且能够互相区分的事物,用矩形表示。实体用一组属性来表示其特性,属性用椭圆来表示。关系指实体间的相互关联,一般用菱形表示,分为一对一、一对多、多对多三种。

任务 2　创建和操作数据库

🖳 任务描述

对网上购物系统进行设计后,要在 MySQL 中创建数据库,并完成相应操作。本任务主要使用图形化管理工具和命令行两种方式实现数据库的创建和操作。

2.2.1　创建数据库

1. 使用图形化管理工具创建数据库

【例 2-1】　使用 Navicat 图形化管理工具,创建名称为"shopping"的数据库。

步骤如下：

启动 Navicat 图形化管理工具,连接服务器,为连接命名为"mysql";右键选择新建数据库命令,输入名称"shopping",字符集选择"utf8mb3--UTF-8 Unicode",排序规则选择"utf8mb3_bin",完成数据库的创建。在界面左侧资源管理器中可以查看名称为"shopping"的数据库。创建过程如图 2-4 至图 2-6 所示。

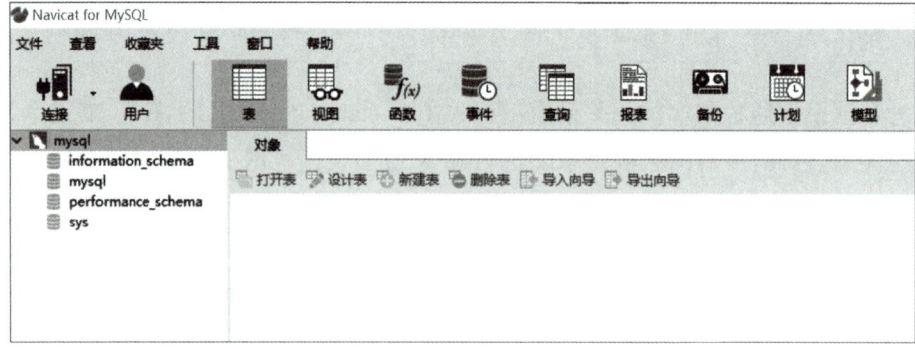

图 2-4　创建数据库 1

图 2-5　创建数据库 2

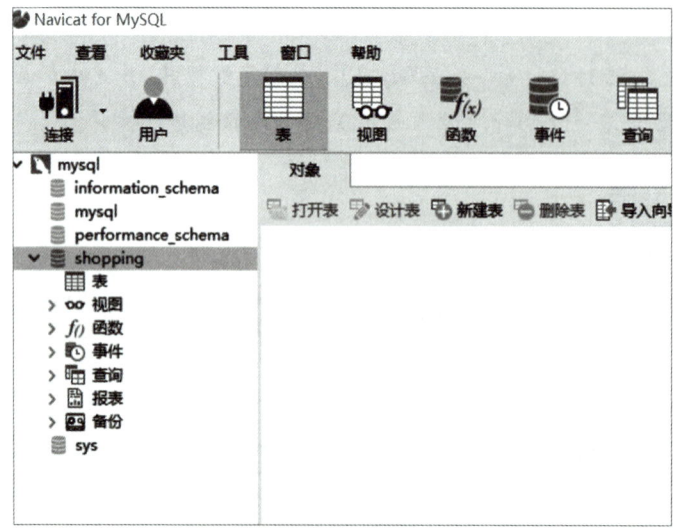

图 2-6　创建数据库 3

2. 使用 SQL 命令语句创建数据库

使用 SQL 命令语句创建数据库，其语法格式为：

CREATE DATABASE 数据库名称

CHARACTER SET 编码方式

COLLATE 排序规则

【例 2-2】 使用 SQL 语句创建名称为“shoppingdb1”的数据库，字符集和排序规则使用默认值。结果显示如图 2-7 所示。

```
MySQL 8.0 Command Line Client
Enter password: *******
Welcome to the MySQL monitor.  Commands end with ; or \g.
Your MySQL connection id is 19
Server version: 8.0.30 MySQL Community Server - GPL

Copyright (c) 2000, 2022, Oracle and/or its affiliates.

Oracle is a registered trademark of Oracle Corporation and/or its
affiliates. Other names may be trademarks of their respective
owners.

Type 'help;' or '\h' for help. Type '\c' to clear the current input statement.

mysql> CREATE DATABASE shoppingdb1;
Query OK, 1 row affected (0.01 sec)
```

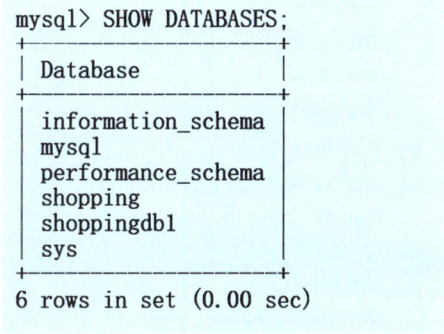

图 2-7　创建“shoppingdb1”数据库

注：“shoppingdb1”指新建数据库的名称，数据库在同一服务器中的名称是唯一的。

2.2.2　查看和选择数据库

1. 查看数据库

【例 2-3】 查看 MySQL 服务器中存在的数据库。结果显示如图 2-8 所示。

```
mysql> SHOW DATABASES;
+--------------------+
| Database           |
+--------------------+
| information_schema |
| mysql              |
| performance_schema |
| shopping           |
| shoppingdb1        |
| sys                |
+--------------------+
6 rows in set (0.00 sec)
```

图 2-8　查看数据库

注：“SHOW DATABASES ”表示查看服务器中的数据库列表语句。

【例 2-4】 查看数据库"shopping"的字符集和排序规则信息。结果显示如图 2-9 所示。

```
mysql> SHOW CREATE DATABASE shopping;
+----------+--------------------------------------------------------------------------------------------------------------------------------+
| Database | Create Database                                                                                                                |
+----------+--------------------------------------------------------------------------------------------------------------------------------+
| shopping | CREATE DATABASE `shopping` /*!40100 DEFAULT CHARACTER SET utf8mb3 COLLATE utf8mb3_bin */ /*!80016 DEFAULT ENCRYPTION='N' */   |
+----------+--------------------------------------------------------------------------------------------------------------------------------+
1 row in set (0.00 sec)
```

图 2-9 查看数据库"shopping"的字符集和排序规则信息

2. 选择数据库

一般数据库系统中有多个数据库,因此在操作数据库对象前要选择一个数据库。

【例 2-5】 在 MySQL 中,使用 USE 语句选择数据库"shopping"。结果显示如图 2-10 所示。

```
mysql> USE shopping;
Database changed
mysql>
```

图 2-10 选择数据库

2.2.3 修改数据库

数据库创建成功后,若要修改数据库信息,有如下两种方式。

1. 使用图形化管理工具修改数据库

【例 2-6】 使用 Navicat 图形化管理工具,修改数据库"shoppingdb1"的属性信息。
步骤如下:

启动 Navicat 图形化管理工具,连接服务器 MySQL,右键单击数据库"shoppingdb1",选择"数据库属性",进行字符集和排序规则的修改,如图 2-11 所示。

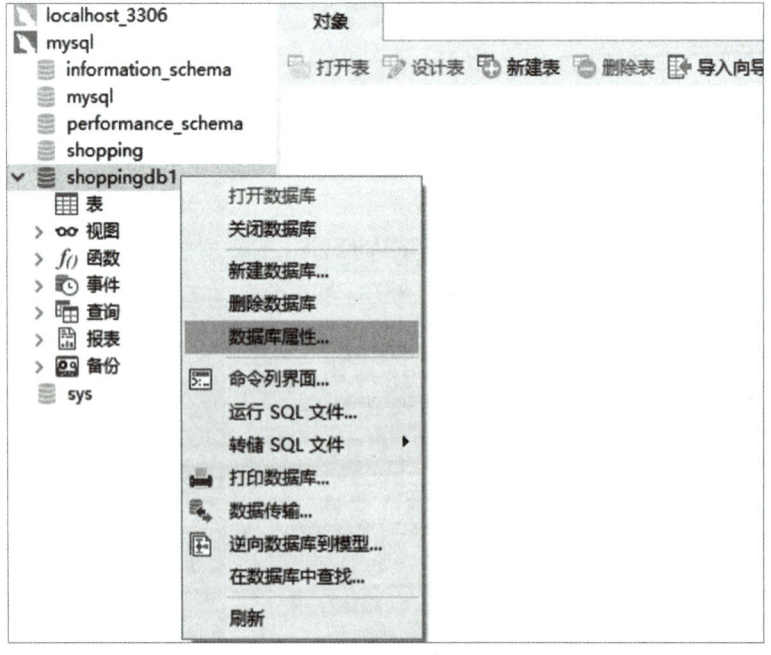

图 2-11 使用 Navicat 图形化管理工具修改数据库

2. 使用 SQL 命令语句修改数据库

使用 SQL 命令语句修改数据库,其语法格式为:

ALTER DATABASE 数据库名称

CHARACTER SET 编码方式

COLLATE 排序规则

【例 2-7】 使用 SQL 命令语句,修改数据库"shopping"的字符集为"gb2312",排序规则为"gb2312_chinese_ci"。结果如图 2-12 所示。

```
mysql> ALTER DATABASE shopping CHARACTER SET gb2312 COLLATE gb2312_chinese_ci;
Query OK, 1 row affected (0.02 sec)
```

图 2-12　使用 SQL 语句修改数据库

修改完成后,查看修改结果。结果如图 2-13 所示。

```
mysql> SHOW CREATE DATABASE shopping;
| Database | Create Database |
| shopping | CREATE DATABASE `shopping` /*!40100 DEFAULT CHARACTER SET gb2312 */ /*!80016 DEFAULT ENCRYPTION='N' */ |
1 row in set (0.00 sec)
```

图 2-13　查看修改结果

由结果可知,数据库"shopping"字符集已经更改为"gb2312"。

2.2.4　删除数据库

数据库删除后,原来占用的空间会释放出来。删除数据库系统中已经存在的数据库有两种方式。

1. 使用图形化管理工具删除数据库

【例 2-8】 使用 Navicat 图形化管理工具,删除数据库"shoppingdb1"。操作如图 2-14 所示。

2. 使用 SQL 命令语句删除数据库

使用 SQL 命令语句删除数据库,其语法格式为:

DROP DATABASE 数据库名称

【例 2-9】 使用 SQL 语句删除名称为"shoppingdb1"的数据库。结果如图 2-15 所示。

删除数据库后,可以使用 SHOW 语句来查看数据库是否删除成功。结果如图 2-16 所示。

从结果看,数据库系统中已经没有名称为"shoppingdb1"的数据库,删除成功。数据库删除后,所有的表和数据都不能恢复,因此,在执行删除操作时要慎重。

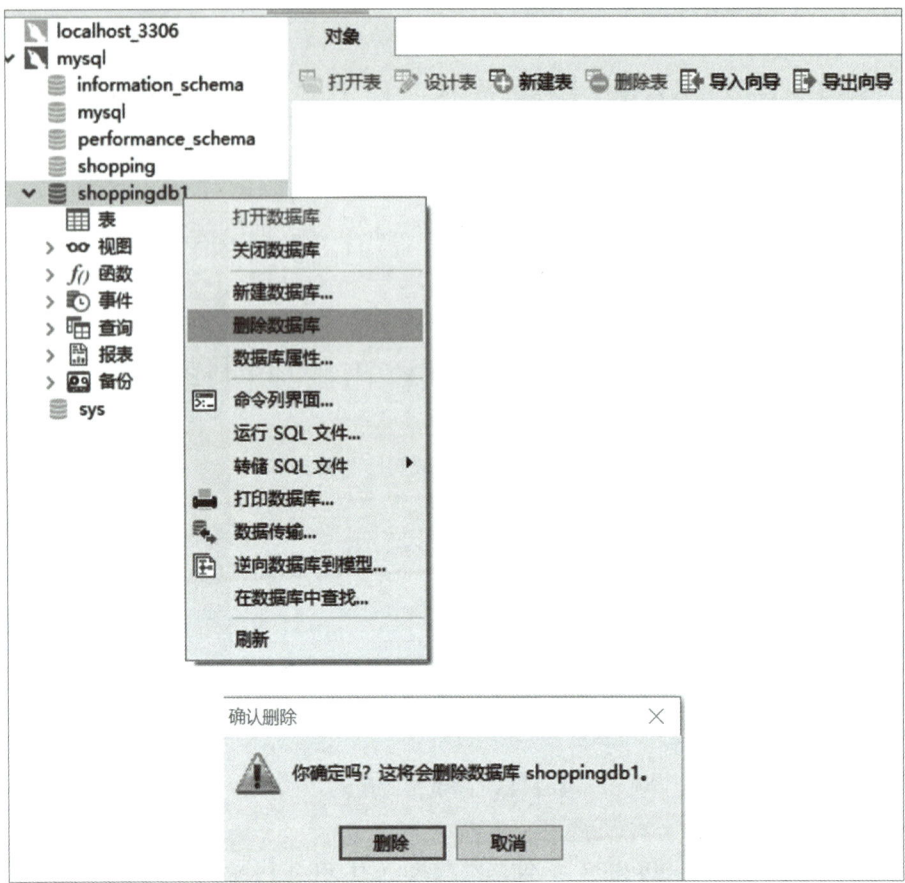

图 2-14　删除数据库 1

```
mysql> DROP DATABASE shoppingdb1;
Query OK, 0 rows affected (0.05 sec)
```

图 2-15　删除数据库 2

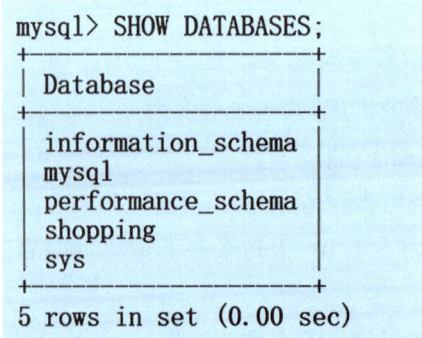

图 2-16　查看数据库是否删除成功

任务3　MySQL 的存储引擎和数据类型

任务描述

在 MySQL 中，数据是存储在数据表中的。在创建数据表时，需要选择存储引擎和数据类型。本任务主要学习 MySQL 的存储引擎和数据类型的相关知识。

2.3.1 MySQL 的存储引擎

存储引擎是关于数据存储的技术，在关系型数据库中，数据是以数据表的形式来存储的，因此存储引擎即为数据表的类型。数据库的存储引擎决定了数据表在计算机中的存储方式，MySQL 数据库提供了多种存储引擎，用户需要选择合适的存储引擎，从而获得较好的整体性能。

1. MySQL 中常用的存储引擎

（1）InnoDB 存储引擎

InnoDB 是 MySQL 的默认存储引擎，也是最重要、应用最广泛的存储引擎，具有提交、回滚和崩溃恢复能力以及多版本并发控制等特性，能够高效处理大量数据，在非事务型存储的需求中，MySQL 一般优先考虑 InnoDB。

（2）MyISAM 存储引擎

MyISAM 存储引擎是 MySQL 5.1 及之前版本的默认引擎，设计简单，提供了大量特性，包括全文索引、压缩、空间函数等，广泛应用于 Web 环境和数据仓库。

（3）Memory 存储引擎

Memory 存储引擎将表中数据存储在内存中，不需要进行磁盘 I/O，查询速度快，主要适用于目标数据较小，且被频繁访问的情况。

2. 查看 MySQL 支持的存储引擎

使用 SQL 语句查询 MySQL 支持的存储引擎，其语法格式为：

SHOW ENGINES

【例 2-10】　查看 MySQL 支持的存储引擎。结果如图 2-17 所示。

使用 SQL 语句可以查询系统默认的存储引擎，其语法格式为：

SHOW VARIABLES LIKE 'default_storage_engine'

【例 2-11】　查看 MySQL 支持的默认存储引擎。结果如图 2-18 所示。

2.3.2 MySQL 的数据类型

在数据库中，数据的表示形式称为数据类型，数据类型决定了数据的存储格式和有效

```
mysql> SHOW ENGINES;
+--------------------+---------+----------------------------------------------------------------+--------------+------+------------+
| Engine             | Support | Comment                                                        | Transactions | XA   | Savepoints |
+--------------------+---------+----------------------------------------------------------------+--------------+------+------------+
| MEMORY             | YES     | Hash based, stored in memory, useful for temporary tables      | NO           | NO   | NO         |
| MRG_MYISAM         | YES     | Collection of identical MyISAM tables                          | NO           | NO   | NO         |
| CSV                | YES     | CSV storage engine                                             | NO           | NO   | NO         |
| FEDERATED          | NO      | Federated MySQL storage engine                                 | NULL         | NULL | NULL       |
| PERFORMANCE_SCHEMA | YES     | Performance Schema                                             | NO           | NO   | NO         |
| MyISAM             | YES     | MyISAM storage engine                                          | NO           | NO   | NO         |
| InnoDB             | DEFAULT | Supports transactions, row-level locking, and foreign keys     | YES          | YES  | YES        |
| BLACKHOLE          | YES     | /dev/null storage engine (anything you write to it disappears) | NO           | NO   | NO         |
| ARCHIVE            | YES     | Archive storage engine                                         | NO           | NO   | NO         |
+--------------------+---------+----------------------------------------------------------------+--------------+------+------------+
9 rows in set (0.00 sec)
```

图 2-17 查看 MySQL 支持的存储引擎

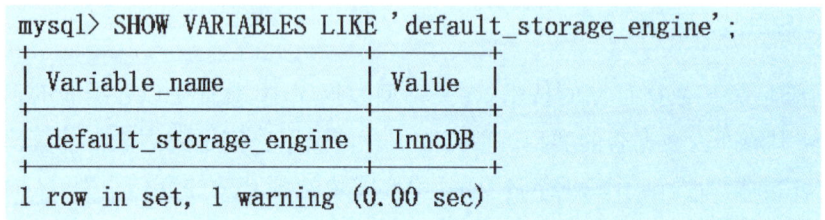

图 2-18 查看 MySQL 支持的默认存储引擎

范围。MySQL 提供了多种数据类型,包括整数类型、小数类型、日期和时间类型、字符串类型、JSON 类型等。

数据类型指数据类型的名称,显示宽度指能够显示的最大数据长度。MySQL 在数据类型名称后指定该类型的显示宽度。如果不指定显示宽度,则使用 MySQL 默认的宽度值。例如,某字段设定数据类型为 int(10),表示该数值最大能够显示的数值是 10 位。要表示数据的取值范围,可在设置数据类型时,加上参数 zerofill,表示零填充,即当数值不足以显示宽度时,用 0 来填补。

1. 整数类型

整数类型是数据库中最基本的数据类型,MySQL 支持的整数类型有 TINYINT、SMALLINT、MEDIUMINT、INT(INTEGER)、BIGINT,其具体类型和取值范围见表 2-5。

表 2-5 MySQL 的整数类型和取值范围

整数类型	字节数(Byte)	无符号数据取值范围	有符号数据取值范围
TINYINT	1	0～255	−128～127
SMALLINT	2	0～65535	−32768～32767
MEDIUMINT	3	0～16777215	−8388608～8388607
INT	4	0～4294967295	−2147483648～2147483647
BIGINT	8	0～18446744073709551615	−9223372036854775808～9223372036854775807

2. 小数类型

在 MySQL 中用浮点数和定点数来表示小数。浮点数在数据库中存放的是近似值，包括单精度浮点类型和多精度浮点类型；定点数在数据库中存放的是精确值。浮点类型和定点类型对应的存储大小和取值范围见表 2-6。

表 2-6　MySQL 的小数类型和取值范围

类型名称	字节数 (Byte)	负数的取值范围	非负数的取值范围
FLOAT	4	$-3.402823466E+38 \sim$ $-1.175494351E-38$	0 和 $1.175494351E-38 \sim$ $3.402823466E+38$
DOUBLE	8	$-1.7976931348623157E+308 \sim$ $-2.2250738585072014E-308$	0 和 $2.2250738585072014E-308 \sim$ $1.7976931348623157E+308$
DECIMAL(M,D) 或 dec(M,D)	$M+2$	与 DOUBLE 相同	与 DOUBLE 相同

3. 日期与时间类型

MySQL 提供了多种表示日期和时间的数据类型，用于存储日期和时间数据，见表 2-7。

表 2-7　MySQL 的日期与时间类型数据

类型名称	字节数 (Byte)	存储格式	存储范围
YEAR	1	YYYY	1901～2155
DATE	3	YYYY-MM-DD	1000-01-01～9999-12-31
TIME	3	HH:MM:SS	-838:59:59～838:59:59
DATETIME	5	YYYY-MM-DD HH:MM:SS	1000-01-01 00:00:00～9999-12-31 23:59:59
	4	YYYY-MM-DD HH:MM:SS	1970-01-01 00:00:01 UTC～2038-01-19 03:14:07 UTC

4. 字符串类型

字符串类型是一种非常重要的数据类型，MySQL 提供了很丰富的字符串类型，BINARY 和 VARBINARY 用于存储较短的二进制字符串，TEXT 和 BLOB 用于存储较大的数据。BINARY 长度固定，VARBINARY 长度可变。TEXT 存储的是文本字符串，例如新闻正文等。BLOB 存储的是二进制字符串，例如图片、音乐等。字符串的具体类型见表 2-8。

表 2-8　MySQL 的字符串类型数据

类型	字节数(Byte)	存储范围
CHAR(M)	$M \times w$	$0 \leqslant M \leqslant 255$
VARCHAR(M)	$L+1$	$0 \leqslant M \leqslant 65\ 535$
BINARY(N)	N	$0 \leqslant N \leqslant 255$
VARBINARY(N)	$L+1$	$0 \leqslant N \leqslant 65\ 535$
BLOB	$L+2$	$L<2^{16}$
TEXT	$L+2$	$L<2^{16}$
ENUM	1 或 2	$0\sim 65\ 535$
SET	1、2、3、4 或 8	最多 64 个成员

注：① w 表示字符集中的最大长度字符所需字节数。
②　L 表示给定字符串值的实际字节长度(以字节为单位)。
③　M 表示非二进制字符串类型的声明列长度(以字符为单位)。
④　N 表示二进制字符串类型的声明列长度(以字符为单位)。

5. JSON 类型

JSON 是自 MySQL 5.7.8 版本开始新增的一种数据类型，用于存储 JSON 数据。其所需空间与 LONGBLOB 和 LONGTEXT 大致相同。JSON 类型字段的插入值可以分为数组和对象。JSON 数组是一个由逗号分隔并包含在"[ ]"中的值列表。例如，["abc",12,NULL,TRUE,FALSE]。JSON 对象是一组键值，由逗号分隔，包含在"{ }"中，例如，{"datebase":"MySQL","language":"Java"}，而且 JSON 数据和 JSON 对象允许嵌套。

任务 4　创建和操作数据表

任务描述

数据表是数据库中最核心的操作对象，一个数据库包含若干个数据表。在关系型数据库中，基础数据都存放在数据表中。本任务主要介绍如何创建数据表、查看数据表，以及如何修改、删除数据表等操作。

2.4.1　创建数据表

1. 使用 Navicat 图形化管理工具创建数据表

【例 2-12】　使用 Navicat 图形化管理工具，在"shopping"数据库中新建会员表，表的名称为"user"，其结构见表 2-9。

表 2-9 会员表(user)结构

序号	字段名	数据类型	标识	主键	允许空	默认值	注释
1	u_id	int	是	是	否		会员 ID
2	u_name	varchar(50)			否		用户名
3	u_pwd	varchar(50)			否		密码
4	u_gender	char(2)			是		性别
5	u_birthday	date			是		出生日期
6	u_credit	int			是	0	积分

操作过程：

（1）打开 Navicat，在资源管理器中双击服务器 MySQL，打开"shopping"数据库，在弹出的操作对象中选择"表"，右键单击"表"，选择新建表命令，如图 2-19 所示。

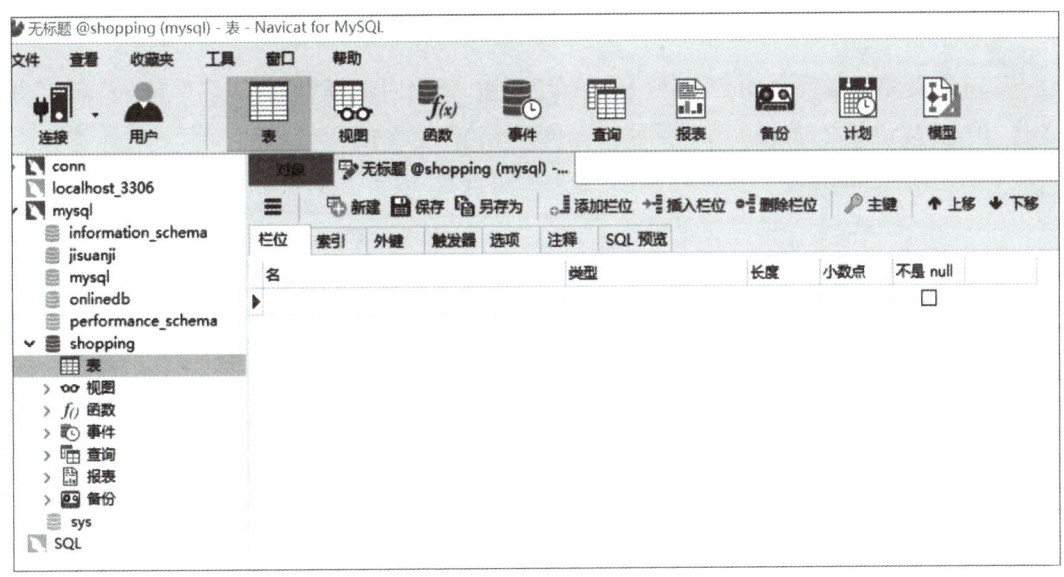

图 2-19 新建表

（2）在表设计窗口，输入字段名、数据类型、长度、小数点、是否允许为空、是否主键，输入注释内容，u_id 列选择"自动递增"复选框，u_credit 列输入默认值 0，如图 2-20 所示。

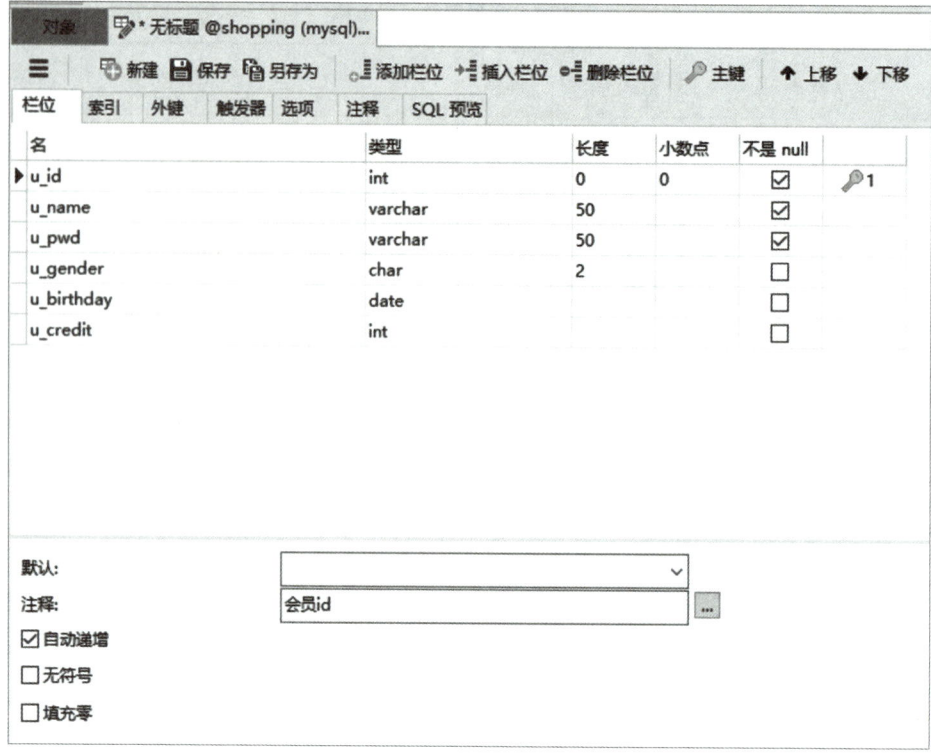

图 2-20 输入

（3）定义完字段后，单击工具栏上"保存"按钮，弹出对话框"输入表名"，输入"user"并单击"确定"按钮即可，如图 2-21 所示。

（4）刷新"shopping"数据库，在表对象栏中可查看"user"表，如图 2-22 所示。

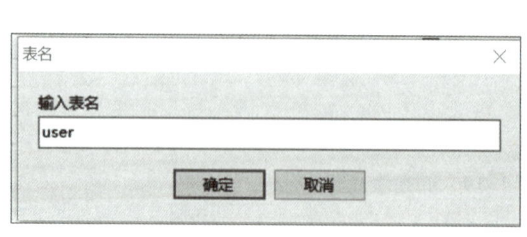

图 2-21 输入表名

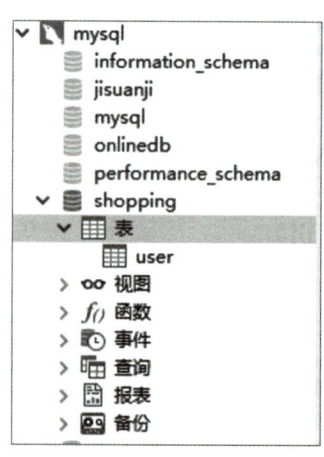

图 2-22 查看"user"表

2. 使用 SQL 语句创建数据表

使用 CREATE TABLE 创建数据表的语法格式为：

CREATE TABLE [数据库库名.]表名

（字段1,

字段2,

……

字段n）

其中,表名指数据表的名称。若不在当前数据库下操作,需要输入数据库库名,其基本形式为"数据库库名.数据表名"。数据表的名称应使用有意义的英文词汇,只能用英文字母、数字、下划线,且要以英文字母开头,词汇间用下划线分隔,不能超过32个字符,不能使用SQL的关键字。定义字段包括字段名、数据类型、长度、是否允许为空,还要指明默认值、主键、注释等。

• AUTO_INCREMENT:将字段设置为自动增长,自动增长量基数为1,步长为1,一个数据表中只能有一个自动增长字段,且只有整数类型才可以设置。

• NULL(NOT NULL):字段是否允许为空。

• DEFAULT:设置字段的默认值。

• PRIMARY KEY:主键约束。

• UNIQUE:唯一性约束。

• COMMENT:注释字段。

【例2-13】 使用SQL语句完成"user"表的创建。结果如图2-23所示。

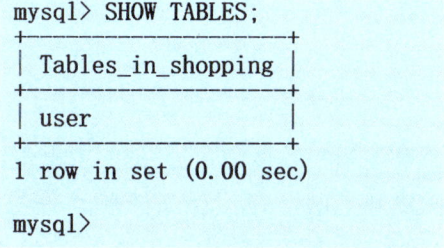

图2-23 "user"表的创建

注:"user"表中共有6个字段,字段与字段间用","隔开,每个字段都有注释,u_id字符类型为int,设置为主键和自动增长字段,u_credit默认值设置为0。

【例2-14】 使用SHOW TABLES语句查看数据表。结果如图2-24所示。

```
mysql> SHOW TABLES;
+-------------------+
| Tables_in_shopping |
+-------------------+
| user              |
+-------------------+
1 row in set (0.00 sec)

mysql>
```

图2-24 查看数据表

3. 查看表结构

在向表中插入数据时,需要先查看表的结构。在 MySQL 中可以用 SHOW CREATE TABLE 和 DESCRIBE 语句进行表结构查询。

(1) 使用 SHOW CREATE TABLE 语句查看表结构,其语法格式为:

SHOW CREATE TABLE 表名

【例 2-15】 使用 SHOW CREATE TABLE 查看"user"表的结构。结果如图 2-25 所示。

```
mysql> SHOW CREATE TABLE user\G;
*************************** 1. row ***************************
       Table: user
Create Table: CREATE TABLE `user` (
  `u_id` int NOT NULL AUTO_INCREMENT COMMENT '会员id',
  `u_name` varchar(50) NOT NULL COMMENT '用户名',
  `u_pwd` varchar(50) NOT NULL COMMENT '密码',
  `u_gender` char(2) DEFAULT NULL COMMENT '性别',
  `u_birthday` date DEFAULT NULL COMMENT '出生日期',
  `u_credit` int DEFAULT '0' COMMENT '积分',
  PRIMARY KEY (`u_id`)
) ENGINE=InnoDB DEFAULT CHARSET=gb2312
1 row in set (0.00 sec)
```

图 2-25 查看"user"表的结构 1

由结果可以看出,SHOW CREATE TABLE 不仅可以查看数据表的定义,还可以查看数据表的存储引擎和字符集。

(2) 使用 DESCRIBE 语句查看"user"表的结构。结果如图 2-26 所示。

```
mysql> DESCRIBE user;
+------------+-------------+------+-----+---------+----------------+
| Field      | Type        | Null | Key | Default | Extra          |
+------------+-------------+------+-----+---------+----------------+
| u_id       | int         | NO   | PRI | NULL    | auto_increment |
| u_name     | varchar(50) | NO   |     | NULL    |                |
| u_pwd      | varchar(50) | NO   |     | NULL    |                |
| u_gender   | char(2)     | YES  |     | NULL    |                |
| u_birthday | date        | YES  |     | NULL    |                |
| u_credit   | int         | YES  |     | 0       |                |
+------------+-------------+------+-----+---------+----------------+
6 rows in set (0.00 sec)
```

图 2-26 查看"user"表的结构 2

2.4.2 设置约束条件

为了保证数据的准确性和逻辑一致性,防止数据库中出现错误数据、无效数据以及不符合语义的数据等,在 MySQL 中,通常设置约束条件来实现数据的完整性。这里的约束条件主要有 PRIMARY KEY 约束、NOT NULL 约束、DEFAULT 约束、UNIQUE 约束、AUTO_INCREMENT 约束和 FOREIGN KEY 约束。

1. PRIMARY KEY 约束

PRIMARY KEY 约束称为主键约束,用于定义数据表中主键一列或多列。主键是唯一标识数据表中的记录,主键字段不能为空且值唯一。主键可以是单一字段,也可以是多个字段的组合。每张表中最多只能有一个主键约束。

【例 2-16】 使用 Navicat 创建商品信息表(goods),其结构见表 2-10。

表 2-10 商品信息表(goods)

序号	字段名	数据类型	主键	允许空	说明
1	g_id	int	是	否	商品编号
2	g_name	varchar(100)	—	否	商品名称
3	g_price	decimal(10,2)	—	是	商品价格
4	g_addtime	datetime	—	否	上架时间

操作过程如下:

(1) 在 Navicat 的"shopping"数据库下新建表。

(2) 在表的设计窗口输入以上定义的表结构。

(3) 选择 g_id 字段,在工具栏中选择主键按钮;或右键单击该字段,在弹出菜单中选择"主键",g_id 最后一列如出现一把钥匙,表示设置成功。结果如图 2-27 所示。

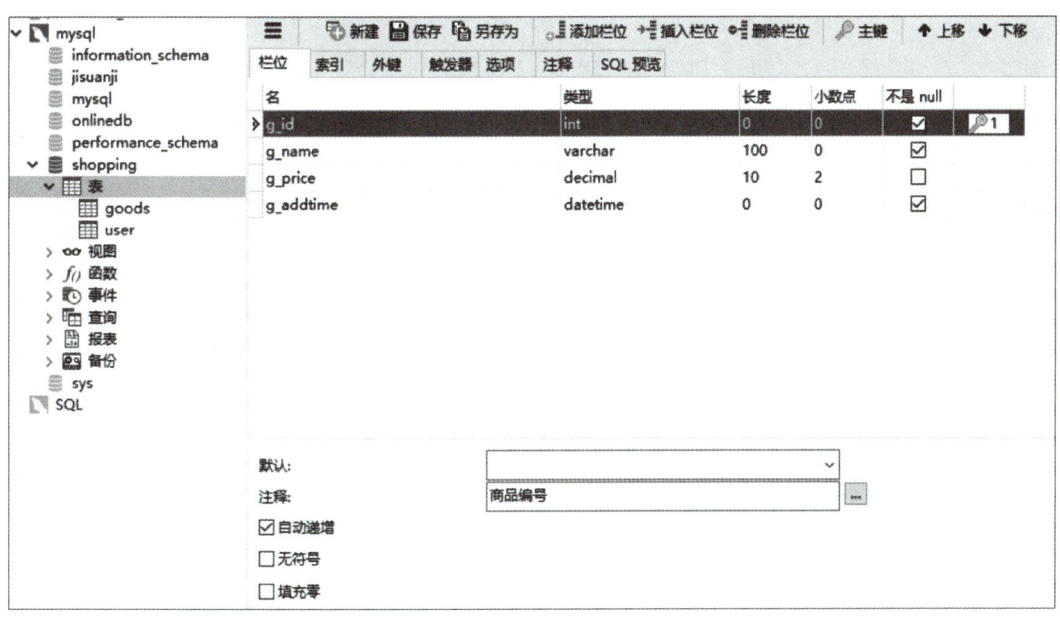

图 2-27 使用 Navicat 创建商品信息表

【例 2-17】 按照例 2-16 中的表结构,使用 SQL 语句,创建商品信息表(goods),并使用 PRIMARY KEY 设置 g_id 为主键约束。结果如图 2-28 所示。

```
mysql> CREATE TABLE goods
    -> (g_id int PRIMARY KEY,
    -> g_name varchar(100) NOT NULL,
    -> g_price decimal(10,2),
    -> g_addtime datetime NOT NULL
    -> );
Query OK, 0 rows affected (0.03 sec)
```

图 2-28 使用 SQL 语句创建商品信息表

若数据表中的主键由多个字段构成,主键要在字段定义后设置,其语法格式为:
PRIMARY KEY(字段 1,字段 2,……,字段 n)

【例 2-18】 创建订单表(orders),其表结构见表 2-11。

表 2-11 订单表(orders)

序号	字段名	数据类型	主键	允许空	说明
1	o_id	int	是	否	订单 ID
2	u_id	int	是	否	用户 ID
3	o_time	datetime	—	否	下单时间
4	o_amount	decimal(10,2)	—	是	订单金额

使用 SQL 语句创建该表。结果如图 2-29 所示。

```
mysql> CREATE TABLE orders
    -> (o_id int,
    -> u_id int,
    -> o_time datetime NOT NULL,
    -> o_amount decimal(10,2),
    -> PRIMARY KEY(o_id,u_id)
    -> );
Query OK, 0 rows affected (0.02 sec)
```

图 2-29 创建订单表

2. NOT NULL 约束

NOT NULL 约束也称为非空约束,设置字段值不能为空,其中 NULL 不同于 0 或空字符,且不能与任何值进行比较,如例 2-17、例 2-18 中 NOT NULL 的使用。通过查看表结构,确定非空约束。

使用 SQL 语句查看"goods"表的结构。结果如图 2-30 所示。

```
mysql> DESC goods;
| Field     | Type          | Null | Key | Default | Extra |
| g_id      | int           | NO   | PRI | NULL    |       |
| g_name    | varchar(100)  | NO   |     | NULL    |       |
| g_price   | decimal(10,2) | YES  |     | NULL    |       |
| g_addtime | datetime      | NO   |     | NULL    |       |
4 rows in set (0.01 sec)
```

图 2-30 查看"goods"表的结构

3. DEFAULT 约束

DEFAULT 约束称为默认值约束,用于指定字段的默认值。当向表中添加数据时,若未为字段赋值,系统会自动将默认值插入。在 SQL 语句中,用关键字 DEFAULT 来标识该约束。其语法格式为:

字段名 数据类型 DEFAULT 默认值

【例 2-19】 创建购物车表"cart",其表结构见表 2-12。

表 2-12 购物车表(cart)

序号	列名	数据类型	标识	允许空	默认值	说明
1	cart_id	int	是	否	—	购物车 ID
2	u_id	int	—	否	—	用户 ID
3	g_id	int	—	否	—	商品 ID
4	c_num	int	—	是	0	购买数量

使用 SQL 语句创建该表。结果如图 2-31 所示。

```
mysql> CREATE TABLE cart
    -> (cart_id int PRIMARY KEY,
    -> u_id int NOT NULL,
    -> g_id int NOT NULL,
    -> c_num int DEFAULT 0
    -> );
Query OK, 0 rows affected (0.20 sec)
```

图 2-31 创建购物车表

4. UNIQUE 约束

UNIQUE 约束称为唯一性约束,指数据表中一个字段或组合字段中只包含唯一值。对字段使用唯一性约束可以防止出现重复数据。主键约束是自动定义为 UNIQUE 约束的,且 UNIQUE 约束可以设置字段值为 NULL。在 SQL 语句中,用关键字 UNIQUE 来标识该约束。其语法格式为:

字段名 数据类型 UNIQUE

【例 2-20】 按照例 2-16 中的表结构,使用 SQL 语句,创建商品信息表(goods_new),并为字段 g_name 添加 UNIQUE 约束。结果如图 2-32 所示。

```
mysql> CREATE TABLE goods_new
    -> (g_id int PRIMARY KEY,
    -> g_name varchar(100) NOT NULL UNIQUE,
    -> g_price decimal(10,2),
    -> g_addtime datetime NOT NULL
    -> );
Query OK, 0 rows affected (0.08 sec)
```

图 2-32 使用 SQL 语句创建商品信息表

5. AUTO_INCREMENT 约束

AUTO_INCREMENT 约束又称为自增约束,指向表中插入数据时,每条记录自动生成编号,并按顺序排列。一张表中只能设置一个字段为自增约束,且该字段是主键,字段类型必须为整数类型。其语法格式为:

字段名 数据类型 AUTO_INCREMENT

如前文例 2-13 所示,这里不再赘述。

6. FOREIGN KEY 约束

FOREIGN KEY 约束又称为外键约束,该约束在两张数据表中实现。在关系型数据库中,数据表之间可以互相关联,一张数据表中的某个字段可能是另一个数据表中的主键。为该字段设置外键约束,可以将两张表关联在一起,保证数据的完整性。两张表必须使用 InnoDB 存储引擎,且设置外键约束的字段和关联的主键必须具有相同的数据类型。

以例 2-12 会员表(user)和例 2-18 订单表(orders)为例,会员表(user)被称为主表,订单表(orders)被称为从表,订单表中的 u_id 称为外键,通过该字段可以实现与主表的关联。

【例 2-21】 使用 Navicat 在订单表(orders)中创建 FOREIGN KEY 约束,约束名称为 FK_orders_u_id。

操作步骤如下:

(1) 在 Navicat 图形化工具中打开订单表(orders)表设计窗口,如图 2-33 所示。

图 2-33 打开表设计窗口

(2) 单击"外键"选项卡,打开设置对话框。输入外键名称"FK_orders_u_id",选择栏位"u_id",参考数据库选择"shopping",参考表选择"user",参考栏位选择"u_id",如图 2-34 所示。

图 2-34 打开设置对话框

(3)单击"保存"按钮,完成表设计。

还可以使用 SQL 语句设置外键约束,其语法格式为:

CONSTRAINT 外键名 FOREIGN KEY(外键字段名) REFERENCES 主表名(主键字段名)

【例 2-22】 (1)使用 SQL 语句,完成例 2-21 的外键设置。结果如图 2-35 所示。

```
mysql> CREATE TABLE orders
    -> (o_id int PRIMARY KEY,
    -> u_id int,
    -> o_time datetime NOT NULL,
    -> o_amount decimal(10,2),
    -> CONSTRAINT FK_orders_u_id FOREIGN KEY(u_id) REFERENCES user(u_id)
    -> );
Query OK, 0 rows affected (0.03 sec)
```

图 2-35 完成外键设置

(4)使用 SHOW CREATE TABLE 查看"orders"表的结构。结果如图 2-36 所示。

```
mysql> SHOW CREATE TABLE orders;
+--------+-------------------+
| Table  | Create Table      |
|        |                   |
+--------+-------------------+
| orders | CREATE TABLE `orders` (
  `o_id` int NOT NULL,
  `u_id` int DEFAULT NULL,
  `o_time` datetime NOT NULL,
  `o_amount` decimal(10,2) DEFAULT NULL,
  PRIMARY KEY (`o_id`),
  KEY `FK_orders_u_id` (`u_id`),
  CONSTRAINT `FK_orders_u_id` FOREIGN KEY (`u_id`) REFERENCES `user` (`u_id`)
) ENGINE=InnoDB DEFAULT CHARSET=gb2312 |
+--------+-------------------+
1 row in set (0.00 sec)
```

图 2-36 查看"orders"表的结构

2.4.3 修改数据表

在系统需求或设计要求变化时,需要对创建完成的数据表的结构进行修改。修改数据表包括修改表名、修改字段名、修改数据类型、增加字段、删除字段和修改字段排列位置等。在 MySQL 中,可以使用图形化管理工具和 SQL 语句来实现修改操作,其中使用图形化管理工具 Navicat 修改数据表的方式与创建数据表的方式相同,这里不再赘述。在 MySQL 中还可使用 ALTER TABLE 语句修改表结构。

1. 修改表名

在 MySQL 中,修改表的名称的关键字是 RENAME,其语法格式为:

ALTER TABLE 原表名 RENAME 新表名

【例 2-23】 将数据库"shopping"中的表"user"更名为"user_new"。结果如图 2-37 所示。

```
mysql> ALTER TABLE user RENAME user_new;
Query OK, 0 rows affected (0.04 sec)
```

图 2-37 修改表名

使用 SHOW TABLES 查看数据库中的表名是否修改成功。结果如图 2-38 所示。

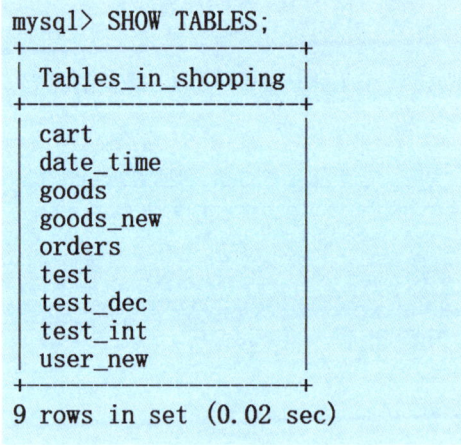

图 2-38 查看表名是否修改成功

2. 修改字段名

在 MySQL 中,修改表的字段名称的关键字是 CHANGE,其语法格式为:

ALTER TABLE 表名 CHANGE 原字段名 新字段名 新数据类型

【例 2-24】 将例 2-23 中的表"user_new"中的"u_name"字段名修改为"u_newname",数据类型为 VARCHAR,长度为 50。结果如图 2-39 所示。

```
mysql> ALTER TABLE user_new CHANGE u_name u_newname varchar(50);
Query OK, 0 rows affected (0.07 sec)
Records: 0  Duplicates: 0  Warnings: 0
```

图 2-39 修改字段名

使用 DESC 语句查看字段名是否修改成功。结果如图 2-40 所示。

```
mysql> DESC user_new;
+------------+-------------+------+-----+---------+----------------+
| Field      | Type        | Null | Key | Default | Extra          |
+------------+-------------+------+-----+---------+----------------+
| u_id       | int         | NO   | PRI | NULL    | auto_increment |
| u_newname  | varchar(50) | YES  |     | NULL    |                |
| u_pwd      | varchar(50) | NO   |     | NULL    |                |
| u_gender   | char(2)     | YES  |     | NULL    |                |
| u_birthday | date        | YES  |     | NULL    |                |
| u_credit   | int         | YES  |     | 0       |                |
+------------+-------------+------+-----+---------+----------------+
6 rows in set (0.02 sec)
```

图 2-40　查看字段名修改结果

3. 修改数据类型

在 MySQL 中，修改字段数据类型的关键字是 MODIFY，其语法格式为：

ALTER TABLE 表名 MODIFY 字段名 新数据类型

【例 2-25】　将例 2-23 中的表"user_new"中"u_birthday"字段的数据类型改为 DATETIME。结果如图 2-41 所示。

```
mysql> ALTER TABLE user_new MODIFY u_birthday datetime;
Query OK, 0 rows affected (0.09 sec)
Records: 0  Duplicates: 0  Warnings: 0
```

图 2-41　修改数据类型

使用 DESC 语句查看数据类型是否修改成功。结果如图 2-42 所示。

```
mysql> DESC user_new;
+------------+-------------+------+-----+---------+----------------+
| Field      | Type        | Null | Key | Default | Extra          |
+------------+-------------+------+-----+---------+----------------+
| u_id       | int         | NO   | PRI | NULL    | auto_increment |
| u_newname  | varchar(50) | YES  |     | NULL    |                |
| u_pwd      | varchar(50) | NO   |     | NULL    |                |
| u_gender   | char(2)     | YES  |     | NULL    |                |
| u_birthday | datetime    | YES  |     | NULL    |                |
| u_credit   | int         | YES  |     | 0       |                |
+------------+-------------+------+-----+---------+----------------+
6 rows in set (0.00 sec)
```

图 2-42　查看数据类型是否修改成功

4. 添加字段

为数据表添加字段有三种类型：①在表的第一列添加字段；②在表的最后一列添加字段；③在表的指定列后添加字段。系统默认在表的最后一列添加字段。在 MySQL 中，添加字段的关键字是 ADD，其语法格式为：

ALTER TABLE 表名 ADD 字段名 数据类型 [FIRST]AFTER 指定字段

使用 DESC 语句查询例 2-16 的"goods"表的结构。结果如图 2-43 所示。

```
mysql> DESC goods;
+-----------+---------------+------+-----+---------+-------+
| Field     | Type          | Null | Key | Default | Extra |
+-----------+---------------+------+-----+---------+-------+
| g_id      | int           | NO   | PRI | NULL    |       |
| g_name    | varchar(100)  | NO   |     | NULL    |       |
| g_price   | decimal(10,2) | YES  |     | NULL    |       |
| g_addtime | datetime      | NO   |     | NULL    |       |
+-----------+---------------+------+-----+---------+-------+
4 rows in set (0.00 sec)
```

图 2-43　查询"goods"表的结构

【例 2-26】　为例 2-16 的"goods"表添加第一列信息，字段名为"g_type"，数据类型为 CHAR(20)，执行并查看表结构。结果如图 2-44 所示。

```
mysql> ALTER TABLE goods ADD g_type char(20) FIRST;
Query OK, 0 rows affected (0.02 sec)
Records: 0  Duplicates: 0  Warnings: 0

mysql> DESC goods;
+-----------+---------------+------+-----+---------+-------+
| Field     | Type          | Null | Key | Default | Extra |
+-----------+---------------+------+-----+---------+-------+
| g_type    | char(20)      | YES  |     | NULL    |       |
| g_id      | int           | NO   | PRI | NULL    |       |
| g_name    | varchar(100)  | NO   |     | NULL    |       |
| g_price   | decimal(10,2) | YES  |     | NULL    |       |
| g_addtime | datetime      | NO   |     | NULL    |       |
+-----------+---------------+------+-----+---------+-------+
5 rows in set (0.00 sec)
```

图 2-44　执行并查看"goods"表的结构 1

【例 2-27】　为"goods"表添加最后一列信息，字段名为"g_num"，数据类型为 INT(10)，执行并查看表结构。结果如图 2-45 所示。

```
mysql> ALTER TABLE goods ADD g_num int(10);
Query OK, 0 rows affected, 1 warning (0.03 sec)
Records: 0  Duplicates: 0  Warnings: 1

mysql> DESC goods;
+-----------+---------------+------+-----+---------+-------+
| Field     | Type          | Null | Key | Default | Extra |
+-----------+---------------+------+-----+---------+-------+
| g_type    | char(20)      | YES  |     | NULL    |       |
| g_id      | int           | NO   | PRI | NULL    |       |
| g_name    | varchar(100)  | NO   |     | NULL    |       |
| g_price   | decimal(10,2) | YES  |     | NULL    |       |
| g_addtime | datetime      | NO   |     | NULL    |       |
| g_num     | int           | YES  |     | NULL    |       |
+-----------+---------------+------+-----+---------+-------+
6 rows in set (0.00 sec)
```

图 2-45　执行并查看"goods"表的结构 2

【例 2-28】 为"goods"表"g_name"字段后添加字段"g_intro",数据类型为 TEXT,执行并查看表结构。结果如图 2-46 所示。

```
mysql> ALTER TABLE goods ADD g_intro TEXT AFTER g_name;
Query OK, 0 rows affected (0.01 sec)
Records: 0  Duplicates: 0  Warnings: 0

mysql> DESC goods;
+-----------+---------------+------+-----+---------+-------+
| Field     | Type          | Null | Key | Default | Extra |
+-----------+---------------+------+-----+---------+-------+
| g_type    | char(20)      | YES  |     | NULL    |       |
| g_id      | int           | NO   | PRI | NULL    |       |
| g_name    | varchar(100)  | NO   |     | NULL    |       |
| g_intro   | text          | YES  |     | NULL    |       |
| g_price   | decimal(10,2) | YES  |     | NULL    |       |
| g_addtime | datetime      | NO   |     | NULL    |       |
| g_num     | int           | YES  |     | NULL    |       |
+-----------+---------------+------+-----+---------+-------+
7 rows in set (0.00 sec)
```

图 2-46 执行并查看"goods"表的结构 3

5. 修改字段排列位置

在 MySQL 中,修改字段顺序的关键字是 MODIFY,其语法格式为:

ALTER TABLE 表名 MODIFY 字段名 1 数据类型 [FIRST]AFTER 字段 2

【例 2-29】 将"goods"表中"g_intro"字段移动到"g_num"字段之后,执行并查看表结构。结果如图 2-47 所示。

```
mysql> ALTER TABLE goods MODIFY g_intro TEXT AFTER g_num;
Query OK, 0 rows affected (0.03 sec)
Records: 0  Duplicates: 0  Warnings: 0

mysql> DESC goods;
+-----------+---------------+------+-----+---------+-------+
| Field     | Type          | Null | Key | Default | Extra |
+-----------+---------------+------+-----+---------+-------+
| g_type    | char(20)      | YES  |     | NULL    |       |
| g_id      | int           | NO   | PRI | NULL    |       |
| g_name    | varchar(100)  | NO   |     | NULL    |       |
| g_price   | decimal(10,2) | YES  |     | NULL    |       |
| g_addtime | datetime      | NO   |     | NULL    |       |
| g_num     | int           | YES  |     | NULL    |       |
| g_intro   | text          | YES  |     | NULL    |       |
+-----------+---------------+------+-----+---------+-------+
7 rows in set (0.00 sec)
```

图 2-47 修改字段排列位置

6. 删除字段

在 MySQL 中,删除表中字段的关键字是 DROP,其语法格式为:

ALTER TABLE 表名 DROP 字段名

【例2-30】 删除"goods"表中的"g_type"字段,执行并查看表结构。结果如图2-48所示。

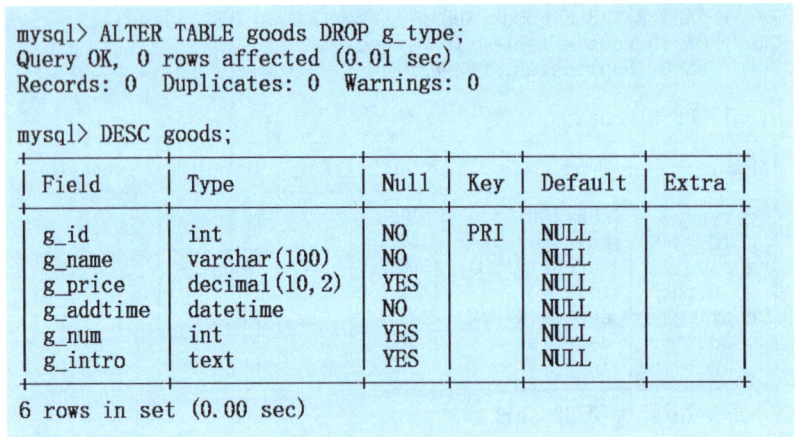

图2-48 删除字段

2.4.4 删除数据表

删除数据表时会将表的结构和表中数据、约束全部删除,因此,用户要反复确认后再执行删除操作。删除数据表时,还要确保数据表中的字段未被其他数据表关联;若有关联,需要先删除关联表。在MySQL中,使用DROP TABLE来删除数据表,其语法格式为:

DROP TABLE 表名

【例2-31】 删除表"test1"。结果如图2-49所示。

```
mysql> DROP TABLE test1;
Query OK, 0 rows affected (0.01 sec)
```

图2-49 删除单张数据表

若要同时删除多张数据表,可以在DROP TABLE后列出表的名称,中间用逗号隔开。

【例2-32】 删除表"test2""test3""test4"。结果如图2-50所示。

```
mysql> DROP TABLE test2,test3,test4;
Query OK, 0 rows affected (0.02 sec)
```

图2-50 删除多张数据表

项目二　操作数据库与数据表

任务5　插入、修改和删除系统数据

任务描述

数据在表中以记录的形式存在,在实际应用中,经常需要对系统数据进行更改,例如:在网上购物系统中,将商品加入购物车、修改购物车里的商品或删除购物车里的商品等。本任务是使用数据操作语句 INSERT、UPDATE、DELETE 实现数据的插入、修改和删除操作。

2.5.1　插入数据

数据插入操作可以向表中添加记录,在 MySQL 中,可以使用 Navicat 图形化管理工具和 SQL 语句两种方法来实现数据插入操作。

1. 使用 Navicat 图形化管理工具插入数据

【例 2-33】　使用 Navicat 为例 2-16 的"goods"表添加商品信息,内容见表 2-13。

表 2-13　商品信息

g_id	g_name	g_price	g_addtime
1	华为 P50 手机	4 258.00	2021-05-01 11:33:40
2	四大名著原著版	450.00	2021-08-10 17:36:07
3	陕西洛川红富士苹果礼盒装	56.20	2022-06-16 12:36:54
4	白象方便面五连包	15.90	2022-07-30 08:37:32

操作步骤:

(1) 打开 Navicat 图形化管理工具中的"shopping"数据库,在表节点下找到"goods",右键选择"打开表",打开数据添加界面,如图 2-51 所示。

(2) 按照上表中的第一行商品信息将数据添加到 Navicat 中,添加完后单击左下方的"+"按钮,新增一行数据,直到将以上四行数据全部录入,录入结束后,单击左下方的"√"按钮即可,如图 2-52 所示。

2. 使用 SQL 语句插入数据

(1) 在 MySQL 中使用 INSERT 语句向表中添加数据,其语法格式为:

INSERT INTO 表名(字段列表) VALUES(值列表)

注:字段列表指需要添加数据的字段名,用括号括起来,不同字段间使用逗号隔开。当向表中所有字段提供值时,字段列表可以省略。值列表指要添加的值,其顺序必须与字段列表顺序一一对应,若字段列表省略时,值列表要按照字段顺序提供值。

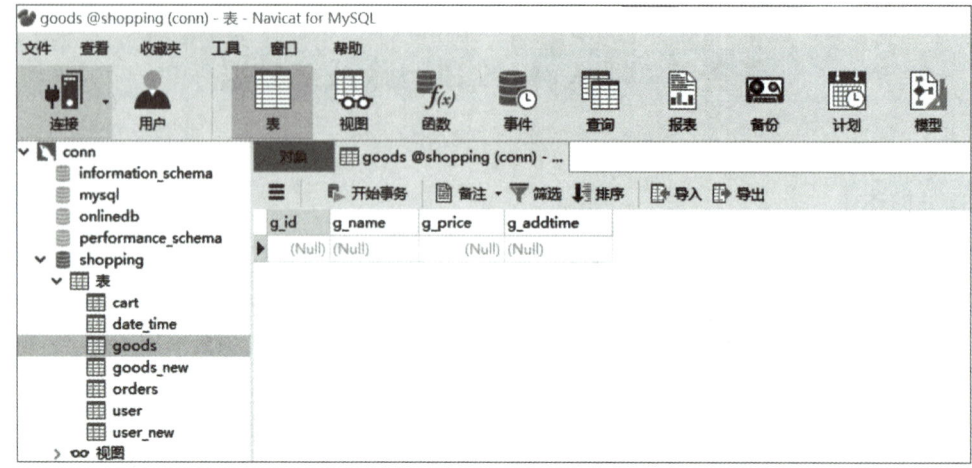

图 2-51　打开数据添加界面

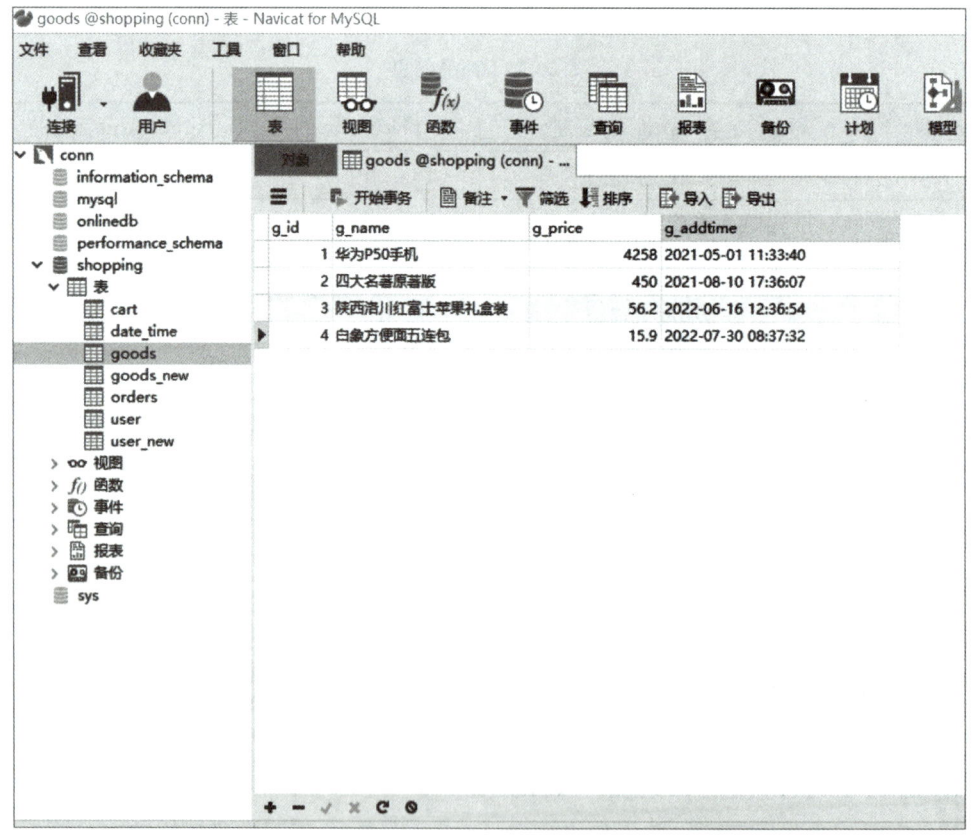

图 2-52　添加数据

【例2-34】 向例2-12中的"user"表插入数据,"u_id"字段值为"1","u_name"字段值为"周明","u_pwd"字段值为"123456","u_gender"字段值为"男"。完成后,查看"user"表中的数据,如图2-53所示。

```
mysql> INSERT INTO user VALUES(1,"周明","123456","男");
Query OK, 1 row affected (0.01 sec)

mysql> SELECT * FROM user;
+------+--------+--------+----------+
| u_id | u_name | u_pwd  | u_gender |
+------+--------+--------+----------+
|    1 | 周明   | 123456 | 男       |
+------+--------+--------+----------+
1 row in set (0.00 sec)
```

图2-53　向"user"表中插入数据

(2) 在MySQL中向表内插入数据时,可以指定一部分字段值,其语法格式为:

INSERT INTO 表名(字段1,字段2,……,字段n) VALUES(字段值1,字段值2,……,字段值n)

【例2-35】 向"goods"表中指定字段插入数据。结果如图2-54所示。

```
mysql> INSERT INTO goods(g_id,g_name,g_addtime) VALUES ('5','耐克男子运动鞋','2022-03-11 11:11:11');
Query OK, 1 row affected (0.00 sec)

mysql> SELECT * FROM goods;
+------+-------------------------+---------+---------------------+
| g_id | g_name                  | g_price | g_addtime           |
+------+-------------------------+---------+---------------------+
|    1 | 华为P50手机              | 4258.00 | 2021-05-01 11:33:40 |
|    2 | 四大名著原著版           |  450.00 | 2021-08-10 17:36:07 |
|    3 | 陕西洛川红富士苹果礼盒装 |   56.20 | 2022-06-16 12:36:54 |
|    4 | 白象方便面五连包         |   15.90 | 2022-07-30 08:37:32 |
|    5 | 耐克男子运动鞋           |    NULL | 2022-03-11 11:11:11 |
+------+-------------------------+---------+---------------------+
5 rows in set (0.00 sec)
```

图2-54　向"goods"表中指定字段插入数据

(3) 在MySQL中,也可以同时向表中添加多条记录,其语法格式为:

INSERT INTO 表名(字段列表) VALUES(值列表1),(值列表2),(值列表3),……,(值列表n)

【例2-36】 向"goods"表中插入多行数据。结果如图2-55所示。

```
mysql> INSERT INTO goods
    -> VALUES(6,'施华洛世奇项链','5378','2021-11-01 12:05:11'),
    -> (7,'小米EA75平板电视','4299','2022-04-10 08:10:35'),
    -> (8,'温碧泉护肤套装','210','2022-07-18 09:20:18');
Query OK, 3 rows affected (0.00 sec)
Records: 3  Duplicates: 0  Warnings: 0
```

图2-55　向"goods"表中插入多行数据

使用 SELECT 语句可以查询"goods"表中数据。结果如图 2-56 所示。

```
mysql> SELECT * FROM goods;
+------+------------------------------+---------+---------------------+
| g_id | g_name                       | g_price | g_addtime           |
+------+------------------------------+---------+---------------------+
|    1 | 华为P50手机                  | 4258.00 | 2021-05-01 11:33:40 |
|    2 | 四大名著原著版               |  450.00 | 2021-08-10 17:36:07 |
|    3 | 陕西洛川红富士苹果礼盒装     |   56.20 | 2022-06-16 12:36:54 |
|    4 | 白象方便面五连包             |   15.90 | 2022-07-30 08:37:32 |
|    5 | 耐克男子运动鞋               |    NULL | 2022-03-11 11:11:11 |
|    6 | 施华洛世奇项链               | 5378.00 | 2021-11-01 12:05:11 |
|    7 | 小米EA75平板电视             | 4299.00 | 2022-04-10 08:10:35 |
|    8 | 温碧泉护肤套装               |  210.00 | 2022-07-18 09:20:18 |
+------+------------------------------+---------+---------------------+
8 rows in set (0.00 sec)
```

图 2-56 查询"goods"表中数据

(4) 在 MySQL 中还可以加入赋值语句,使用 INSERT 为数据表添加数据。其语法格式为:

INSERT INTO 表名 SET 字段名 1=字段值 1,字段名 2=字段值 2,……,字段名 n=字段值 n

【例 2-37】 向"goods"表中插入数据,其中"g_id"为"9","g_name"为"联想笔记本电脑","g_price"为"3699","g_addtime"为"2022-02-25 10:19:28"。结果如图 2-57 所示。

```
mysql> INSERT INTO goods
    -> SET g_id='9',g_name='联想笔记本电脑',g_price='3699',g_addtime='2022-02-25 10:19:28';
Query OK, 1 row affected (0.02 sec)

mysql> SELECT * FROM goods;
+------+------------------------------+---------+---------------------+
| g_id | g_name                       | g_price | g_addtime           |
+------+------------------------------+---------+---------------------+
|    1 | 华为P50手机                  | 4258.00 | 2021-05-01 11:33:40 |
|    2 | 四大名著原著版               |  450.00 | 2021-08-10 17:36:07 |
|    3 | 陕西洛川红富士苹果礼盒装     |   56.20 | 2022-06-16 12:36:54 |
|    4 | 白象方便面五连包             |   15.90 | 2022-07-30 08:37:32 |
|    5 | 耐克男子运动鞋               |    NULL | 2022-03-11 11:11:11 |
|    6 | 施华洛世奇项链               | 5378.00 | 2021-11-01 12:05:11 |
|    7 | 小米EA75平板电视             | 4299.00 | 2022-04-10 08:10:35 |
|    8 | 温碧泉护肤套装               |  210.00 | 2022-07-18 09:20:18 |
|    9 | 联想笔记本电脑               | 3699.00 | 2022-02-25 10:19:28 |
+------+------------------------------+---------+---------------------+
9 rows in set (0.00 sec)
```

图 2-57 向"goods"表中插入数据

2.5.2 修改数据

在实际应用中,用户插入的数据可能出现错误,或随着时间推移,表中数据需要更新。在 MySQL 中,可以使用 Navicat 图形化管理工具和 SQL 语句两种方式修改数据。

(1) 使用 Navicat 图形化管理工具修改数据。打开要修改的数据表,直接进行编辑输入,完成后单击左下角的"√"按钮即可。以"goods"表为例,如图 2-58 所示。

图 2-58　使用 Navicat 图形化管理工具修改数据

（2）在 MySQL 中使用 UPDATE 语句修改表中数据，既可以修改一行也可以修改多行。若无条件表达式，数据表中所有数据都会受到影响；若加入条件表达式，则修改指定条件的数据。其语法格式为：

UPDATE 表名 SET 字段名1＝字段值1,字段名2＝字段值2,……,字段名n＝字段值n

WHERE 条件表达式

【例 2-38】　修改"goods"表，将表中所有商品的"g_addtime"统一修改为"2022-09-01 12:00:00"。结果如图 2-59 所示。

【例 2-39】　修改"goods"表，将表中"g_id"为"5"的"耐克男子运动鞋"修改为"耐克男子跑步鞋"。结果如图 2-60 所示。

2.5.3　删除数据

当不再需要数据表中某些数据时，可以使用 Navicat 图形化管理工具和 SQL 语句两种方法来删除数据。

图 2-59　修改数据 1

图 2-60　修改数据 2

1. 使用 Navicat 图形化管理工具删除数据

在 Navicat 中打开需要进行删除操作的数据表,选中需要删除的行,单击左下角的"—"按钮,弹出"确认删除"提示,选择"删除一条记录"即可。以"goods"表为例,如图 2-61 所示。

2. 使用 SQL 语句删除数据

在 MySQL 中使用 DELETE 语句删除数据,可以删除表中所有数据,也可以删除指定数据,其语法格式为:

DELETE FROM 表名

WHERE 条件表达式

【例 2-40】　通过 SELECT 语句查看"goods_new"表。结果如图 2-62 所示。

图 2-61　删除数据

```
mysql> SELECT * FROM goods_new;
+------+----------------------+---------+---------------------+
| g_id | g_name               | g_price | g_addtime           |
+------+----------------------+---------+---------------------+
|    1 | 华为P50手机          | 4258.00 | 2021-05-01 11:33:40 |
|    2 | 四大名著原著版       |  450.00 | 2021-08-10 17:36:07 |
|    3 | 陕西洛川红富士苹果礼盒装 | 56.20 | 2022-06-16 12:36:54 |
|    4 | 白象方便面五连包     |   15.90 | 2022-07-30 08:37:32 |
|    5 | 耐克男子跑步鞋       |  699.00 | 2022-03-11 11:11:11 |
|    6 | 施华洛世奇项链       | 5378.00 | 2021-11-01 12:05:11 |
|    7 | 小米EA75平板电视     | 4299.00 | 2022-04-10 08:10:35 |
|    8 | 温碧泉护肤套装       |  210.00 | 2022-07-18 09:20:18 |
|    9 | 联想笔记本电脑       | 3699.00 | 2022-02-25 10:19:28 |
+------+----------------------+---------+---------------------+
9 rows in set (0.00 sec)
```

图 2-62　查看"goods_new"表

删除"goods_new"表中的所有数据,执行结果如图 2-63 所示。

```
mysql> DELETE FROM goods_new;
Query OK, 9 rows affected (0.00 sec)

mysql> SELECT * FROM goods_new;
Empty set (0.00 sec)
```

图 2-63　删除"goods_new"表中所有数据

【例2-41】 删除"goods"表中"g_id">4的数据。结果如图2-64所示。

```
mysql> DELETE FROM goods WHERE g_id>4;
Query OK, 5 rows affected (0.00 sec)

mysql> SELECT * FROM goods;
+------+----------------------------+---------+---------------------+
| g_id | g_name                     | g_price | g_addtime           |
+------+----------------------------+---------+---------------------+
|    1 | 华为P50手机                 | 4258.00 | 2021-05-01 11:33:40 |
|    2 | 四大名著原著版               |  450.00 | 2021-08-10 17:36:07 |
|    3 | 陕西洛川红富士苹果礼盒装     |   56.20 | 2022-06-16 12:36:54 |
|    4 | 白象方便面五连包             |   15.90 | 2022-07-30 08:37:32 |
+------+----------------------------+---------+---------------------+
4 rows in set (0.00 sec)
```

图 2-64 删除"g_id">4 的数据

课后习题

一、选择题

1. 创建数据表时,某字段如果不允许其值为空,可以使用( )约束来实现。

 A. NOT NULL   B. NOT BLANK   C. NO NULL   D. NO BLANK

2. 在 MySQL 中,删除数据表中某个字段使用的 SQL 语句是( )。

 A. ALTER TABLE……DELETE……

 B. ALTER TABLE……DELETE COLUMN

 C. ALTER TABLE……DROP……

 D. ALTER TABLE……DROP COLUMN

3. 下列哪个选项不会导致数据输入无效?( )

 A. 列值的存储空间   B. 列的精度

 C. 列的取值范围   D. 使用者的习惯

4. 下列不属于数据操作语言的 SQL 关键字是( )。

 A. SELECT   B. DROP   C. UPDATE   D. DELETE

5. 修改用户信息表中名字为"zhangxia"的数据,将其密码修改为"123456",正确的语句是( )。

 A. UPDATE 用户信息表 SET 密码=123456;

 B. UPDATE 用户信息表 SET 密码=123456 WHERE 名字='zhangxia';

 C. UPDATE SET 密码=123456;

 D. UPDATE 用户信息表 密码=123456 WHERE 管理员名称='zhangxia';

6. 设置表的默认字符集的关键字是( )。

 A. DEFAULT CHARACTER   B. DEFAULT SET

C. DEFAULT   D. DEFAULT CHARACTER SET

7. 下列哪种类型不是 MySQL 中常用的数据类型？（　　）

A. INT　　　　B. VAR　　　　C. CHAR　　　　D. TIME

8. 创建表时，设置某列的值唯一可以使用约束是（　　）。

A. NOT NULL　　B. DEFAULT　　C. CHECK　　D. UNIQUE

二、简答题

1. 在图书管理系统中，图书的借阅时间应该使用哪种数据类型来表达呢？请说明理由。

2. 对于数值类型和字符串类型，空值 NULL 和 0 相同吗？请说明理由。

项目实训

1. 实训任务

（1）用 Navicat 图形化管理工具和 SQL 语句创建和管理数据库。

（2）用 Navicat 图形化管理工具和 SQL 语句创建和管理数据表。

（3）用 Navicat 图形化管理工具和 SQL 语句维护数据的完整性。

（4）用 Navicat 图形化管理工具和 SQL 语句添加、修改和删除系统数据。

2. 实训目的

（1）会使用 Navicat 图形化管理工具创建、管理数据库和数据表。

（2）会使用 SQL 语句创建、管理数据库和数据表。

（3）能根据实际需求，为数据表选择合适的数据类型和约束。

（4）会使用 Navicat 图形化管理工具进行系统数据添加、修改和删除。

（5）会使用 SQL 语句进行系统数据添加、修改和删除。

3. 实训内容

（1）创建名称为"shopping"的数据库，默认字符集设置为 UTF-8mb4。

（2）设置商品信息表（goods）、用户信息表（users）、订单信息表（orders）的表结构，并添加 PRIMARY KEY、UNIQUE、DEFAULT 等约束。

（3）向各数据表中插入有效信息。

（4）在 Navicat 和编程环境中添加、修改和删除信息。

项目三

查询系统数据

数据查询是数据库管理系统最重要的功能之一,也是最基本的操作,可以满足用户对数据的查询、计算、统计和分析等要求。在 MySQL 中,使用 SELECT 语句可对数据进行查询,该语句功能强大,使用方式灵活。本项目通过查询单表数据、查询多表数据等,介绍如何使用 SELECT 语句实现数据查询功能。

学习目标

- 了解基本查询语句
- 会使用 SELECT 语句查询数据列
- 会使用条件语句筛选指定数据行
- 会使用聚合函数进行数据分组统计
- 会使用内连接、外连接、复合条件连接、子查询等查询多表数据

素质目标

引导学生进行有效、合法的数据库查询,正确分析和解决问题,激发学生的学习热情,培养学生勇于探索的创新精神和严谨务实的科学态度。

项目三 查询系统数据

任务 1　查询单表数据

任务描述

单表数据查询指从一张数据表中查询所需要的数据,是最基本的数据查询。本任务主要学习基本查询语句,从而实现在数据表中进行数据列和数据行的查询操作。

3.1.1　SELECT 基本查询语句

在 MySQL 中查询数据的基本语句为 SELECT 语句,其基本语法格式为:
SELECT * |列名 1 AS 别名 1,列名 2 AS 别名 2,……,列名 n AS 别名 n
FROM 表名
【WHERE 条件表达式】
【GROUP BY 列名】
【ORDER BY 列名 ASC|DESC】
注:
SELECT 指查询结果返回的列,使用"*"则可以显示表中所有的列。
FROM 指明要查询的数据表。
WHERE 限定查询行要满足的条件。
GROUP BY 按照指定的字段对查询出来的数据进行分组。
ORDER BY 指定查询结果集的排序,其中 ASC 表示升序,DESC 表示降序。

3.1.2　查询数据列

查询数据列指从表中选出由指定列组成的查询结果集,查询方式有两种:一种是使用通配符"*",一种是列出所需列的字段名。

1. 查询所有字段

使用通配符"*"表示选择表中所有数据列,查询结果会与数据源表顺序相同。

【例 3-1】　查询"shopping"数据库中"goods"表中的所有商品信息。结果如图 3-1 所示。

提示:USE 数据库名称,指选择数据库。如无特殊说明,则操作都在"shopping"数据库中进行。

2. 查询指定字段

使用 SELECT 语句可以查询指定列,列与列之间用逗号隔开。当查询表中所有信息时,也可以将表中字段名全部列出,按照要求为字段列调整结果集顺序。

```
mysql> USE shopping;
Database changed
mysql> SELECT * FROM goods;
+------+--------------------------------+---------+---------------------+
| g_id | g_name                         | g_price | g_addtime           |
+------+--------------------------------+---------+---------------------+
|    1 | 华为P50手机                     | 4258.00 | 2021-05-01 11:33:40 |
|    2 | 四大名著原著版                   |  450.00 | 2021-08-10 17:36:07 |
|    3 | 陕西洛川红富士苹果礼盒装         |   56.20 | 2022-06-16 12:36:54 |
|    4 | 白象方便面五连包                 |   15.90 | 2022-07-30 08:37:32 |
+------+--------------------------------+---------+---------------------+
4 rows in set (0.00 sec)
```

图 3-1　查询所有商品信息

【例 3-2】　查询"shopping"数据库中"goods"表所有商品的"g_name"和"g_price"字段信息。结果如图 3-2 所示。

```
mysql> SELECT g_name,g_price FROM goods;
+--------------------------------+---------+
| g_name                         | g_price |
+--------------------------------+---------+
| 华为P50手机                     | 4258.00 |
| 四大名著原著版                   |  450.00 |
| 陕西洛川红富士苹果礼盒装         |   56.20 |
| 白象方便面五连包                 |   15.90 |
+--------------------------------+---------+
4 rows in set (0.00 sec)
```

图 3-2　查询特定字段信息

【例 3-3】　查询"goods"表中所有信息,并将商品价格调整到结果集最后一列。结果如图 3-3 所示。

```
mysql> SELECT g_id,g_name,g_addtime,g_price FROM goods;
+------+--------------------------------+---------------------+---------+
| g_id | g_name                         | g_addtime           | g_price |
+------+--------------------------------+---------------------+---------+
|    1 | 华为P50手机                     | 2021-05-01 11:33:40 | 4258.00 |
|    2 | 四大名著原著版                   | 2021-08-10 17:36:07 |  450.00 |
|    3 | 陕西洛川红富士苹果礼盒装         | 2022-06-16 12:36:54 |   56.20 |
|    4 | 白象方便面五连包                 | 2022-07-30 08:37:32 |   15.90 |
+------+--------------------------------+---------------------+---------+
4 rows in set (0.00 sec)
```

图 3-3　查询表中所有信息并调整

3. 为结果集指定列名

一般情况下,查询结果集显示的是数据源表中的列标题,若需要为结果集中的列重新定义列名,可以使用关键字 AS。

【例 3-4】 查询"goods"表中商品信息,将列名分别重命名为"商品编号""商品名称""商品价格""上架时间"。结果如图 3-4 所示。

```
mysql> SELECT g_id AS 商品编号,g_name AS 商品名称,g_price AS 商品价格,g_addtime AS 上架时间 FROM goods;
+----------+------------------------+----------+---------------------+
| 商品编号 | 商品名称               | 商品价格 | 上架时间            |
+----------+------------------------+----------+---------------------+
|        1 | 华为P50手机            |  4258.00 | 2021-05-01 11:33:40 |
|        2 | 四大名著原著版         |   450.00 | 2021-08-10 17:36:07 |
|        3 | 陕西洛川红富士苹果礼盒装 |    56.20 | 2022-06-16 12:36:54 |
|        4 | 白象方便面五连包       |    15.90 | 2022-07-30 08:37:32 |
+----------+------------------------+----------+---------------------+
4 rows in set (0.00 sec)
```

图 3-4 列名重命名

3.1.3 查询数据行

在实际应用中,若只需要获取满足用户需求的数据,可以使用 WHERE 子句对数据表进行筛选,其语法格式为:

SELECT 列名 FROM 表名 WHERE 条件表达式

该条件表达式是通过运算符将列名、常量、变量、函数组合起来,运算符包括比较运算符、逻辑运算符、BETWEEN AND 运算符、IN 运算符、LIKE 运算符等。此外,查询数据行还会用到 DISTINCT、LIMIT 等语句。

1. 比较运算符

使用比较运算符可以限定查询条件,语法格式为:

WHERE 表达式 1 比较运算符 表达式 2

常用的比较运算符见表 3-1。

表 3-1 比较运算符

比较运算符	说明	比较运算符	说明
=	相等	!=	不等于
<	小于	>=	大于等于
>	大于	<=	小于等于

【例 3-5】 查询"goods"表中"g_id"=2 的商品信息。结果如图 3-5 所示。

```
mysql> SELECT * FROM goods WHERE g_id=2;
+------+----------------+---------+---------------------+
| g_id | g_name         | g_price | g_addtime           |
+------+----------------+---------+---------------------+
|    2 | 四大名著原著版 |  450.00 | 2021-08-10 17:36:07 |
+------+----------------+---------+---------------------+
1 row in set (0.00 sec)
```

图 3-5 查询"g_id"=2 的商品信息

【例3-6】 查询"goods"表中商品价格在100元以上的商品名称(name)、商品价格(price)和上架时间(addtime)。结果如图3-6所示。

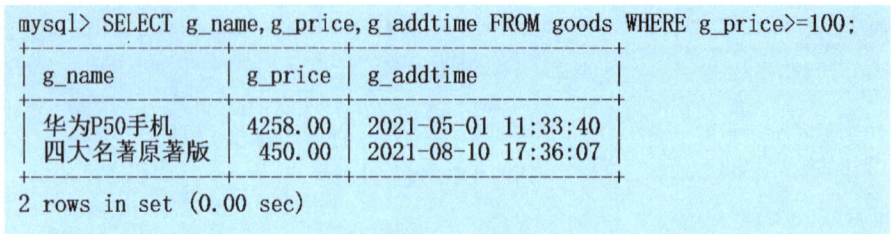

图3-6 查询商品价格等信息

2. 逻辑运算符

逻辑表达式通过逻辑运算符连接起来,逻辑运算结果通常有三种,分别为 TRUE(真)、FALSE(假)、NULL(不确定)。常用的逻辑运算符见表3-2。

表3-2 逻辑运算符

逻辑运算符	说明
逻辑与 AND(&&)	操作数都为真,结果为真,否则为假
逻辑或 OR(||)	操作数都为假,结果为假,否则为真
逻辑非 NOT(!)	操作数为真,结果为假,反之亦然
逻辑异或 XOR	操作数逻辑相反,结果为真;逻辑相同,结果为假

【例3-7】 查询"goods"表中价格在100元以上、5 000元以下的商品信息。结果如图3-7所示。

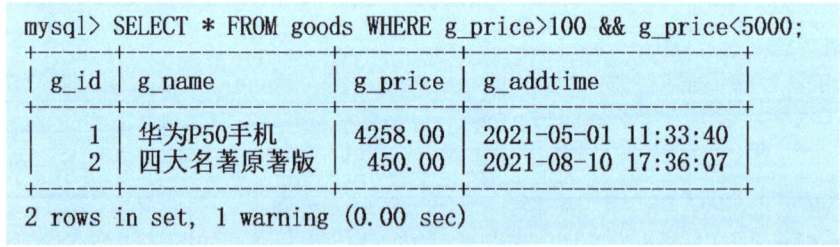

图3-7 查询特定价格商品信息

【例3-8】 查询"goods"表中价格大于50元或数量在30个以下的商品的编号和名称。结果如图3-8所示。

【例3-9】 查询"goods"表中在7月或8月上架的商品名称、商品价格和上架时间。结果如图3-9所示。

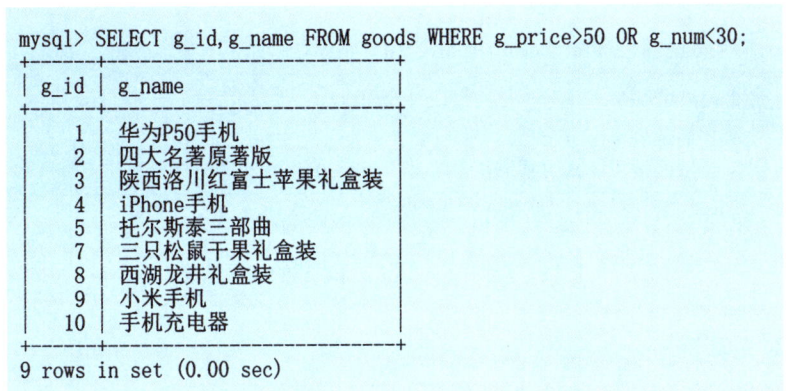

图 3-8　查询特定价格和数量的商品的编号和名称

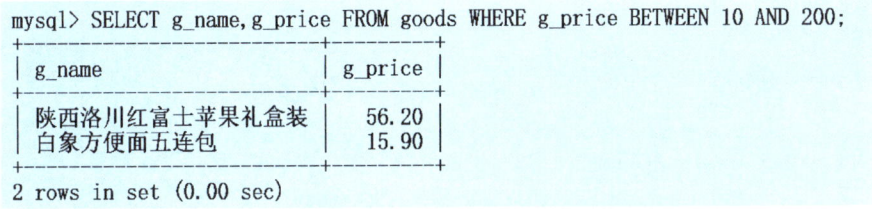

图 3-9　查询特定月份上架的商品信息

3. BETWEEN AND 运算符

在 MySQL 中,使用 BETWEEN AND 查询某个范围内的数据,查询结果包含开始值和结束值。其语法格式为:

WHERE 表达式 BETWEEN 开始值 AND 结束值

【例 3-10】　查询"goods"表中"g_price"的值在 10 到 200 之间的商品名称和商品价格。结果如图 3-10 所示。

```
mysql> SELECT g_name,g_price FROM goods WHERE g_price BETWEEN 10 AND 200;
+------------------------------+---------+
| g_name                       | g_price |
+------------------------------+---------+
| 陕西洛川红富士苹果礼盒装     |   56.20 |
| 白象方便面五连包             |   15.90 |
+------------------------------+---------+
2 rows in set (0.00 sec)
```

图 3-10　查询商品信息

4. IN 运算符

IN 运算符也是用来限制数据查询范围的,其语法格式为:

WHERE 表达式 IN(值 1,值 2,……,值 n)

【例 3-11】　查询"goods"表中"g_id"是"2"和"3"的商品信息。结果如图 3-11 所示。

```
mysql> SELECT * FROM goods WHERE g_id IN(2,3);
+------+----------------------------------+---------+---------------------+
| g_id | g_name                           | g_price | g_addtime           |
+------+----------------------------------+---------+---------------------+
|    2 | 四大名著原著版                     |  450.00 | 2021-08-10 17:36:07 |
|    3 | 陕西洛川红富士苹果礼盒装            |   56.20 | 2022-06-16 12:36:54 |
+------+----------------------------------+---------+---------------------+
2 rows in set (0.00 sec)
```

图 3-11 查询"g_id"是"2"和"3"的商品信息

5. LIKE 运算符

当查询条件只有部分信息时,可以使用 LIKE 运算符进行模糊查询,其语法格式为:
WHERE 列名 LIKE 查询条件

这里与 LIKE 运算符同时使用的字符称为通配符,通常有两种:"％"和"_"。通配符"％"可以匹配任意长度的字符,包括 0 个、1 个或多个。通配符"_"使用方法与"％"类似,但只能匹配 1 个字符。

先向例 2-26 中的"goods"表添加信息。结果如图 3-12 所示。

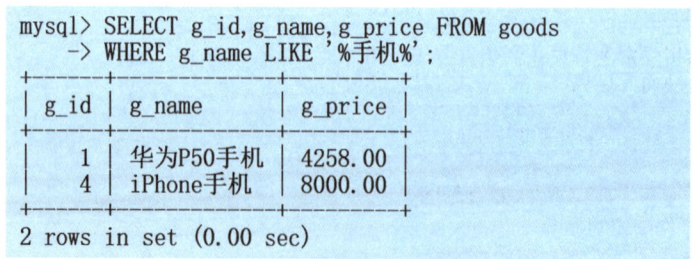

图 3-12 添加信息

【例 3-12】 查询"goods"表中"g_name"中含有"手机"的商品编号、商品名称和商品价格。结果如图 3-13 所示。

```
mysql> SELECT g_id,g_name,g_price FROM goods
    -> WHERE g_name LIKE '%手机%';
+------+-------------+---------+
| g_id | g_name      | g_price |
+------+-------------+---------+
|    1 | 华为P50手机  | 4258.00 |
|    4 | iPhone手机  | 8000.00 |
+------+-------------+---------+
2 rows in set (0.00 sec)
```

图 3-13 模糊查询商品信息 1

【例 3-13】 查询"goods"表中"g_name"第二个字是"尔"的商品名称、商品价格和上架时间。结果如图 3-14 所示。

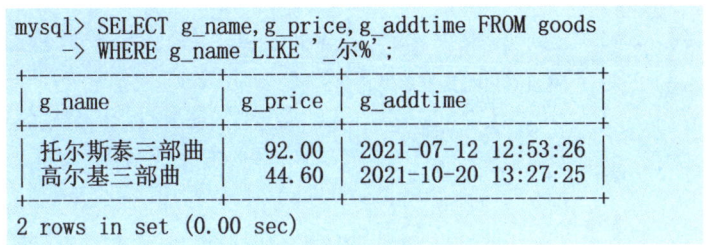

图 3-14 模糊查询商品信息 2

6. DISTINCT 消除重复行

当查询结果集出现重复的记录时,MySQL 提供了 DISTINCT 关键字来消除重复行。

【例 3-14】 查询"goods"表,列出商品种类。结果如图 3-15 所示。

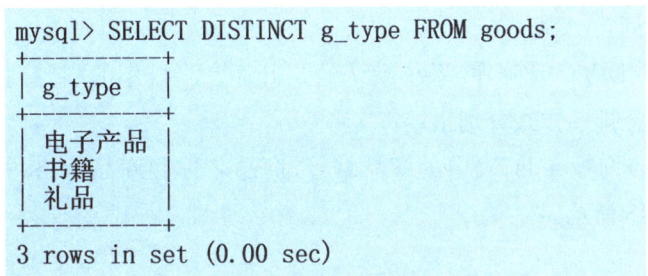

图 3-15 列出商品种类

7. LIMIT 限制返回行数

在实际查询中,有时只需要返回满足条件的部分记录,这样既便于阅读查看,又避免系统资源浪费。在 MySQL 中,关键字 LIMIT 可以限制返回的记录数,还可以指定查询结果从哪一条开始。

【例 3-15】 查询"goods"表,返回前 3 行结果集。结果如图 3-16 所示。

```
mysql> SELECT * FROM goods LIMIT 3;
+--------+------+---------------------------+-------+---------+---------------------+
| g_type | g_id | g_name                    | g_num | g_price | g_addtime           |
+--------+------+---------------------------+-------+---------+---------------------+
| 电子产品 |    1 | 华为P50手机                |    10 | 4258.00 | 2021-05-01 11:33:40 |
| 书籍    |    2 | 四大名著原著版              |    30 |  450.00 | 2021-08-10 17:36:07 |
| 礼品    |    3 | 陕西洛川红富士苹果礼盒装     |    40 |   56.20 | 2022-06-16 12:36:54 |
+--------+------+---------------------------+-------+---------+---------------------+
3 rows in set (0.00 sec)
```

图 3-16 返回满足条件的记录

【例 3-16】 查询"goods"表,列出第 3~5 行的商品编号、商品名称和商品数量。结果如图 3-17 所示。

```
mysql> SELECT g_id,g_name,g_num FROM goods LIMIT 2,3;
+------+----------------------------+-------+
| g_id | g_name                     | g_num |
+------+----------------------------+-------+
|    3 | 陕西洛川红富士苹果礼盒装   |    40 |
|    4 | iPhone手机                 |    25 |
|    5 | 托尔斯泰三部曲             |    50 |
+------+----------------------------+-------+
3 rows in set (0.00 sec)
```

图 3-17　返回指定行数的记录

3.1.4　数据排序

在实际应用中,除了对数据进行基本查询外,还需要对数据做排序、统计等操作。默认情况下结果集是按照数据源中数据的物理顺序排列的。在 MySQL 中使用 ORDER BY 实现对数据的排序,其语法格式为:

ORDER BY 列名 ASC|DESC

其中,ASC 是升序,DESC 是降序,系统默认的是升序。当排序的指定列不止一列时,列名之间用逗号隔开,且排序方式分别指定。

【例 3-17】　查询"goods"表中的商品编号、商品名称和商品价格,并按照价格降序排列。结果如图 3-18 所示。

```
mysql> SELECT g_id,g_name,g_price FROM goods ORDER BY g_price DESC;
+------+----------------------------+---------+
| g_id | g_name                     | g_price |
+------+----------------------------+---------+
|    4 | iPhone手机                 | 8000.00 |
|    1 | 华为P50手机                | 4258.00 |
|    2 | 四大名著原著版             |  450.00 |
|    5 | 托尔斯泰三部曲             |   92.00 |
|    3 | 陕西洛川红富士苹果礼盒装   |   56.20 |
|    6 | 高尔基三部曲               |   44.60 |
+------+----------------------------+---------+
6 rows in set (0.00 sec)
```

图 3-18　价格降序排列

【例 3-18】　查询"goods"表中的商品名称、商品价格和商品数量,按照商品数量升序排序,数量相同的按照商品价格降序排列。结果如图 3-19 所示。

```
mysql> SELECT g_name,g_price,g_num FROM goods ORDER BY g_num ASC,g_price DESC;
+----------------------------+---------+-------+
| g_name                     | g_price | g_num |
+----------------------------+---------+-------+
| 华为P50手机                | 4258.00 |    10 |
| iPhone手机                 | 8000.00 |    20 |
| 四大名著原著版             |  450.00 |    20 |
| 托尔斯泰三部曲             |   92.00 |    40 |
| 陕西洛川红富士苹果礼盒装   |   56.20 |    40 |
| 高尔基三部曲               |   44.60 |    70 |
+----------------------------+---------+-------+
6 rows in set (0.00 sec)
```

图 3-19　数量、价格排序

3.1.5 数据分组统计

对表进行数据查询后,通常需要对查询结果进行统计分析,例如,需要统计商品数量,商品价格最大值、最小值等。在 MySQL 中,使用聚合函数和 GROUP BY 子句对查询结果集进行分组和统计。

1. 聚合函数

聚合函数可以实现对数据表中指定的列值进行统计计算,MySQL 提供的常用聚合函数有以下五种:SUM()函数、AVG()函数、MAX()函数、MIN()函数和 COUNT()函数。

(1) SUM()函数:用于计算指定字段的总和,语法格式为:

SUM(列名)

【例 3-19】 查询"goods"表,计算表中所有商品的数量和,列名设为"总量"。结果如图 3-20 所示。

```
mysql> SELECT SUM(g_num) AS 总量 FROM goods;
+------+
| 总量 |
+------+
|  180 |
+------+
1 row in set (0.00 sec)
```

图 3-20 求数量和

(2) AVG()函数:用于计算指定字段的平均值。AVG()函数常与 GROUP BY 一起使用,可以计算分组数据的平均值。语法格式为:

AVG(列名)

【例 3-20】 查询"goods"表,计算每个商品类别的平均价格。结果如图 3-21 所示。

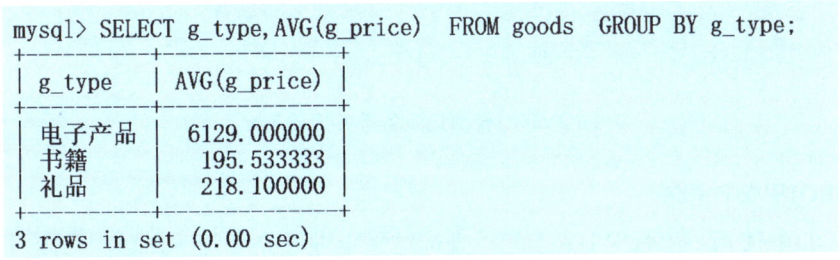

图 3-21 计算平均价格

(3) MAX()函数和 MIN()函数:用于求指定字段中的"最大值"和"最小值",语法格式为:

MAX(列名)、MIN(列名)

【例 3-21】 查询"goods"表中商品的最高价格和最低价格,列名设为最高价和最低价。结果如图 3-22 所示。

```
mysql> SELECT MAX(g_price)AS 最高价,MIN(g_price)AS 最低价 FROM goods;
+---------+--------+
| 最高价  | 最低价 |
+---------+--------+
| 8000.00 | 44.60  |
+---------+--------+
1 row in set (0.00 sec)
```

图 3-22　查询最高和最低价格

（4）COUNT()函数：用于统计表中符合条件的记录的条数,语法格式有两种：COUNT(*)指计算表中所有记录的条数；COUNT(列名)指计算表中指定字段记录的条数,空值忽略不计。

【例 3-22】 查询"goods"表中商品的记录条数。结果如图 3-23 所示。

```
mysql> SELECT COUNT(*) FROM goods;
+----------+
| COUNT(*) |
+----------+
|        7 |
+----------+
1 row in set (0.01 sec)
```

图 3-23　查询记录条数

【例 3-23】 查询"goods"表中有库存的商品记录条数,列名设置为"库存商品"。结果如图 3-24 所示。

```
mysql> SELECT COUNT(g_num) AS 库存商品 FROM goods;
+----------+
| 库存商品 |
+----------+
|        5 |
+----------+
1 row in set (0.00 sec)
```

图 3-24　查询记录条数并设置列名

2. GROUP BY 子句

MySQL 中使用 GROUP BY 子句对查询结果集进行分组,且 GROUP BY 子句通常与聚合函数一起使用。GROUP BY 子句还可以和 GROUP_CONCAT 函数一起使用,将分组中指定列的列值显示出来。另外,GROUP BY 子句和 HAVING 子句一起使用,可以限定结果集满足的条件。其语法格式为：

GROUP BY 列名

HAVING 条件表达式

【例3-24】 统计"goods"表中每种类别(g_type)的商品的数量。结果如图3-25所示。

```
mysql> SELECT g_type,COUNT(g_id) FROM goods GROUP BY g_type;
+-----------+-------------+
| g_type    | COUNT(g_id) |
+-----------+-------------+
| 电子产品  |           2 |
| 书籍      |           3 |
| 礼品      |           2 |
+-----------+-------------+
3 rows in set (0.00 sec)
```

图3-25 统计每种类别的商品数量

【例3-25】 将"goods"表按照类别(g_type)分组,并显示每组中的商品名称。结果如图3-26所示。

```
mysql> SELECT g_type,GROUP_CONCAT(g_name) FROM goods GROUP BY g_type;
+-----------+----------------------------------------------------+
| g_type    | GROUP_CONCAT(g_name)                               |
+-----------+----------------------------------------------------+
| 书籍      | 四大名著原著版,托尔斯泰三部曲,高尔基三部曲         |
| 电子产品  | 华为P50手机,iPhone手机                             |
| 礼品      | 陕西洛川红富士苹果礼盒装,三只松鼠干果礼盒装        |
+-----------+----------------------------------------------------+
3 rows in set (0.00 sec)
```

图3-26 按类别分组并显示名称

【例3-26】 将"goods"表按照类别(g_type)分组,统计每组的数量,并显示商品数量大于2的商品类别。结果如图3-27所示。

```
mysql> SELECT g_type,COUNT(g_id) FROM goods GROUP BY g_type HAVING COUNT(g_id)>2;
+-----------+-------------+
| g_type    | COUNT(g_id) |
+-----------+-------------+
| 书籍      |           3 |
| 礼品      |           3 |
+-----------+-------------+
2 rows in set (0.00 sec)
```

图3-27 分组统计

3.1.6 使用图形化管理工具实现单表查询

【例3-27】 使用Navicat图形化管理工具查询"goods"表中商品价格高于100元的商品类型、商品名称、商品价格,并按照商品价格降序排列,操作如下:

(1) 打开Navicat工具,选择包含数据表的数据库,单击"查询"按钮,选择"新建"→"查询创建工具"。结果如图3-28所示。

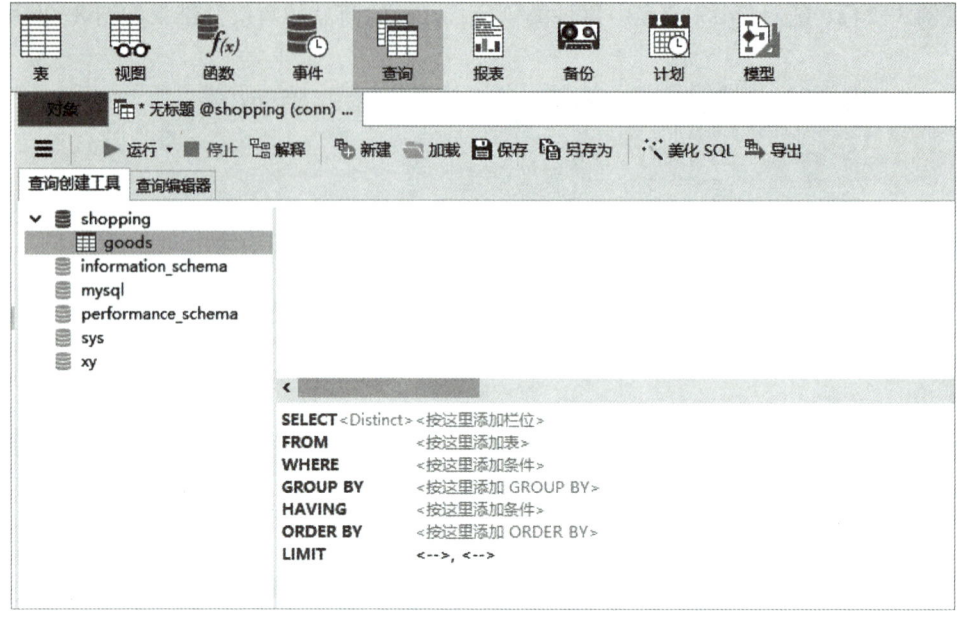

图 3-28　新建查询

（2）在下方的编辑区域找到"按这里添加表"，选择需要查询的数据表。单击"按这里添加栏位"，选择需要查询的列，也可以在"编辑"中直接填写查询内容。结果如图 3-29 所示。

图 3-29　表查询

(3)单击"WHERE"后的"按这里添加条件",选择"g_price>100",单击"ORDER BY"后的"按这里添加条件",选择"g_price DESC"。结果如图 3-30 所示。

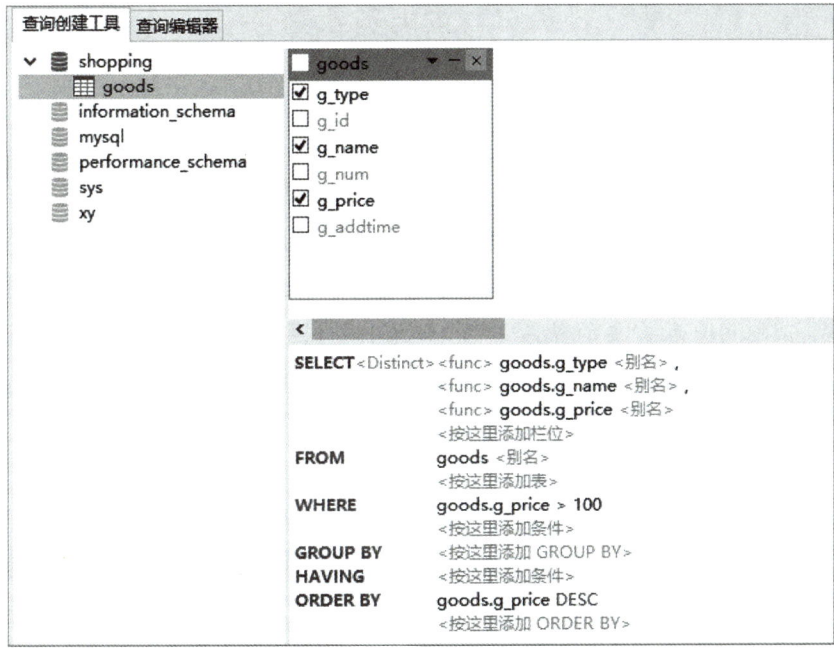

图 3-30　添加条件表查询

(4)单击选项卡中的"运行"按钮,查询语句和结果集都会在"查询编辑器"选项卡中显示,如图 3-31 所示。

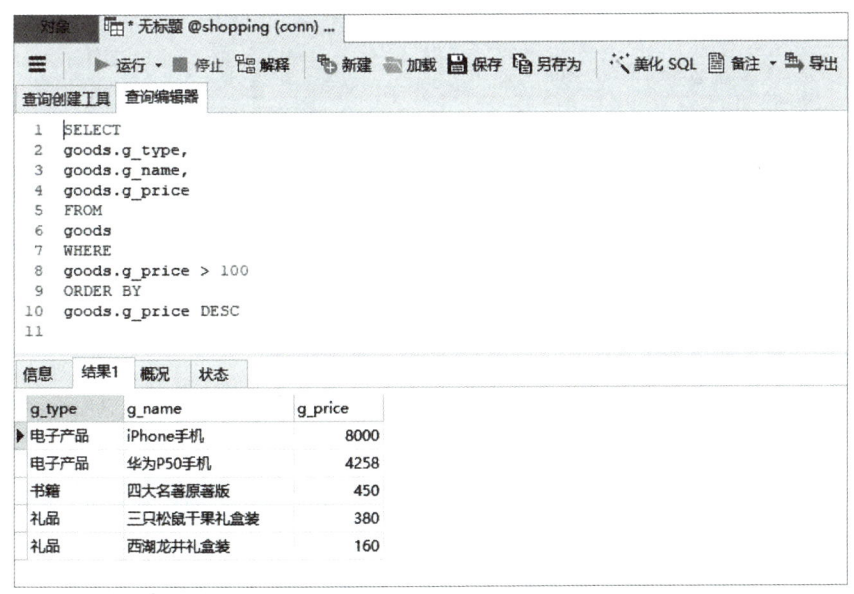

图 3-31　显示查询语句和结果

任务 2 查询多表数据

任务描述

在实际应用中,查询经常会涉及互相关联的多张数据表。本任务主要掌握多表数据查询方法,包括内连接查询、外连接查询、复合条件连接查询和子查询。

连接查询是数据库中非常重要的一种查询类型,通过表与表之间的相关列进行多张表之间的关联,获取用户所需要的信息。在 MySQL 中,通过 JOIN 关键字来实现连接查询操作。

3.2.1 内连接查询

内连接查询(INNER JOIN)是数据表连接查询最常用的操作,一般使用比较运算符对多个表之间的数据进行比较,列出与连接条件匹配的数据记录。表与表之间的连接条件由两个表共有的列组成。内连接查询的基本语法格式为:

SELECT 列名 FROM 表 1 名称

INNER JOIN 表 2 名称 ON 连接条件

【例 3-28】 查询商品表和订单表,列出售出商品的 g_id、名称、销售金额和下单时间。结果如图 3-32 所示。

```
mysql> SELECT goods.g_id,g_name,o_amount,o_addtime
    -> FROM goods JOIN orders
    -> ON goods.g_id=orders.g_id;
+------+--------------------------------+----------+---------------------+
| g_id | g_name                         | o_amount | o_addtime           |
+------+--------------------------------+----------+---------------------+
|    1 | 华为P50手机                    |  8307.00 | 2021-08-30 11:57:50 |
|    2 | 四大名著原著版                 |  1200.00 | 2022-06-10 09:30:24 |
|    3 | 陕西洛川红富士苹果礼盒装       |   560.00 | 2022-06-19 19:36:03 |
|    4 | iPhone手机                     | 16000.00 | 2022-02-06 10:37:27 |
+------+--------------------------------+----------+---------------------+
4 rows in set (0.00 sec)
```

图 3-32 查询商品表和订单表

3.2.2 外连接查询

内连接查询返回符合连接条件的记录,但在实际应用中,经常需要显示某个表的全部行,即使这些行不满足连接条件。例如,查询所有用户的订单金额,即使该用户未下订单也要显示。这种情况下,查询结果集中除了显示符合连接条件的记录外,还要显示至少一个表中的所有记录,不满足条件的显示"NULL"。这叫做外连接,可以分为左外连接和右外连接。

1. 左外连接

查询结果中除包含满足连接条件的记录外，还包括左表中不满足条件的记录，左表中不满足条件的记录在右表中对应的列值为"NULL"。其基本语法格式为：

SELECT 列名 FROM 表1名称 LEFT JOIN 表2名称

ON 表1.共有列名＝表2.共有列名

2. 右外连接

查询结果中除包含满足连接条件的记录外，还包括右表中不满足条件的记录，右表中不满足条件的记录在左表中对应的列值为"NULL"。其基本语法格式为：

SELECT 列名 FROM 表1名称 RIGHT JOIN 表2名称

ON 表1.共有列名＝表2.共有列名

【例3-29】 用左外连接查询用户信息表和订单表，列出用户 u_id、用户名和订单金额。结果如图 3-33 所示。

```
mysql> SELECT users.u_id,u_name,o_amount
    -> FROM users LEFT JOIN orders
    -> ON users.u_id=orders.u_id;
+------+--------+----------+
| u_id | u_name | o_amount |
+------+--------+----------+
|    1 | 李向阳 |  8307.00 |
|    2 | 张霞   |  1200.00 |
|    3 | 刘晓宇 |   560.00 |
|    4 | 周丽   | 16000.00 |
|    5 | 李雅萍 |     NULL |
|    6 | 张辉   |     NULL |
+------+--------+----------+
6 rows in set (0.00 sec)
```

图 3-33　左外连接查询信息

【例3-30】 用右外连接查询用户信息表和订单表，列出每个用户的 u_id、用户名和订单金额。结果如图 3-34 所示。

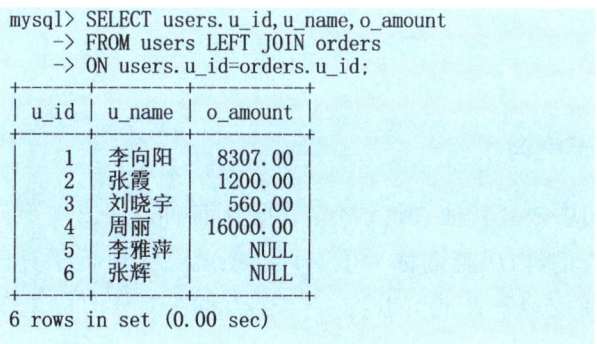

图 3-34　右外连接查询信息

3.2.3 复合条件连接查询

复合条件连接查询是指在连接查询中添加筛选条件,与单表查询类似。

【例 3-31】 查询商品表和订单表,列出售出电子产品的 g_id、名称、销售金额和下单时间。结果如图 3-35 所示。

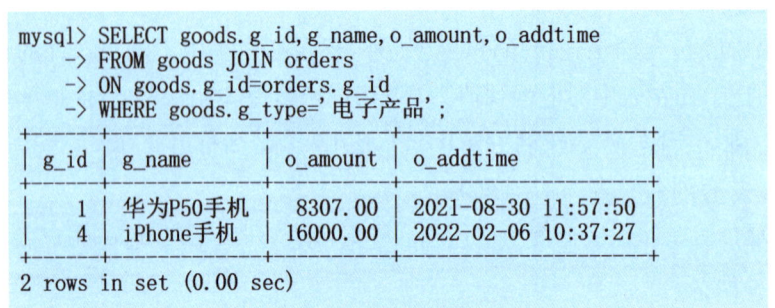

图 3-35 复合条件查询

3.2.4 主查询与子查询

当一个查询语句中嵌套其他查询语句,数据查询的条件依赖于其他的查询结果时,将外层查询称为主查询,内层查询称为子查询。系统会先执行子查询,将子查询的结果作为主查询的条件使用。子查询可以将一个复杂的查询分解成单个查询命令来解决。本小节主要介绍 WHERE 子句中的子查询。

先执行子查询语句,并将返回的结果作为主查询的筛选条件。在 WHERE 子句中的子查询通常使用比较运算符,IN 运算符,ANY、SOME、ALL 谓词和 EXISTS 等关键字。

➢ 比较运算符:当子查询返回的结果集为单值时,可以使用比较运算符为外层查询提供比较操作。

➢ IN 运算符:当子查询返回的结果集是一个集合,主查询需要返回符合条件的集合中的记录时,可以使用 IN 关键字。

➢ ANY、SOME、ALL 谓词:当子查询返回的结果集为单值时,可以使用 ANY、SOME、ALL 等谓词对表达式和子查询结果进行比较,得到结果集。

➢ EXISTS:系统使用 EXISTS 对子查询的返回结果进行判断,若子查询有至少一行记录,则结果为 TRUE,执行主查询;若子查询没有任何记录返回,则结果为 FALSE,主查询不执行。

【例 3-32】 查询用户名是"张霞"的订单 ID、订单金额和下单时间。结果如图 3-36 所示。

【例 3-33】 查询商品信息表中销售金额超过 2 000 元的商品的 ID、名称和上架时间。结果如图 3-37 所示。

```
mysql> SELECT o_id,o_amount,o_addtime
    -> FROM orders
    -> WHERE u_id=(SELECT u_id FROM users WHERE u_name='张霞');
+------+----------+---------------------+
| o_id | o_amount | o_addtime           |
+------+----------+---------------------+
|    2 |  1200.00 | 2022-06-10 09:30:24 |
+------+----------+---------------------+
1 row in set (0.00 sec)
```

图 3-36　查询订单

```
mysql> SELECT goods.g_id,g_name,g_addtime FROM goods
    -> WHERE g_id IN(SELECT g_id FROM orders WHERE o_amount>'2000');
+------+-----------+---------------------+
| g_id | g_name    | g_addtime           |
+------+-----------+---------------------+
|    1 | 华为P50手机 | 2021-05-01 11:33:40 |
|    4 | iPhone手机  | 2022-01-01 15:32:34 |
+------+-----------+---------------------+
2 rows in set (0.00 sec)
```

图 3-37　查询商品信息

【例 3-34】　查询商品信息表中 2022 年 1 月 1 日以后下单的商品的信息。结果如图 3-38 所示。

```
mysql> SELECT * FROM goods WHERE g_id=ANY(SELECT g_id FROM orders
    -> WHERE o_addtime>'2022-01-01 00:00:00');
+----------+------+------------------+-------+---------+---------------------+
| g_type   | g_id | g_name           | g_num | g_price | g_addtime           |
+----------+------+------------------+-------+---------+---------------------+
| 书籍     |    2 | 四大名著原著版    |    20 |  450.00 | 2021-08-10 17:36:07 |
| 礼品     |    3 | 陕西洛川红富士苹果礼盒装 | 40 |   56.20 | 2022-06-16 12:36:54 |
| 电子产品 |    4 | iPhone手机       |  NULL | 8000.00 | 2022-01-01 15:32:34 |
| 礼品     |    8 | 西湖龙井礼盒装    |  NULL |  160.00 | 2022-05-30 15:07:08 |
+----------+------+------------------+-------+---------+---------------------+
4 rows in set (0.00 sec)
```

图 3-38　查询特定时间商品信息

【例 3-35】　查询用户信息表中下过订单的用户的 ID、用户名和生日。结果如图 3-39 所示。

```
mysql> SELECT u_id,u_name,u_birthday
    -> FROM users
    -> WHERE EXISTS (SELECT * FROM orders WHERE u_id=users.u_id);
+------+--------+---------------------+
| u_id | u_name | u_birthday          |
+------+--------+---------------------+
|    1 | 李向阳  | 1990-07-22 11:53:51 |
|    2 | 张霞    | 1993-08-30 08:54:30 |
|    3 | 刘晓宇  | 2000-02-11 15:55:24 |
|    4 | 周丽    | 1998-11-01 02:30:07 |
+------+--------+---------------------+
4 rows in set (0.00 sec)
```

图 3-39　查询用户信息

当数据更新需要依赖于某个查询结果集时,使用子查询更加有效。

(1) 插入查询结果集

在开发测试过程中有需要复制表的情况,例如将一个表内满足条件的数据复制到另一张表中,可以使用 INSERT……SELECT 语句,这种方式比使用单行 INSERT 语句效率高。

其语法格式为:

INSERT [INTO] 表名(列名1,列名2,……,列名n)

SELECT 列名1,列名2……

FROM 表名

WHERE 条件表达式

【例3-36】 创建热销商品表(goods_hot),将日销售量大于等于30的商品定为热销款,将其添加到热销商品表中。结果如图3-40所示。

```
mysql> INSERT INTO goods_hot
    -> SELECT * FROM goods
    -> WHERE g_num>=30;
Query OK, 2 rows affected, 2 warnings (0.03 sec)
Records: 2  Duplicates: 0  Warnings: 2
```

图 3-40　插入查询结果集

(2) 子查询用于修改数据

当数据的更改需要用到其他查询结果时,可以使用子查询作为更新条件。其语法格式为:

UPDATE 表名

SET 字段名

WHERE 子查询

【例3-37】 为购物车表中意向购买数量超过5的用户赠送100积分,结果如图3-41所示。

```
mysql> UPDATE users
    -> SET u_credit=u_credit+10
    -> WHERE u_id IN(SELECT u_id
    -> FROM cart
    -> WHERE c_num>=5);
Query OK, 3 rows affected (0.03 sec)
Rows matched: 3  Changed: 3  Warnings: 0
```

图 3-41　修改数据

(3)子查询用于删除数据

当数据的删除需要用到其他查询结果时,可以使用子查询作为删除条件。其语法格式为:

DELETE FROM 表名

WHERE 子查询

【例3-38】 从商品信息表中删除热销商品信息。结果如图3-42所示。

```
mysql> DELETE FROM goods
    -> WHERE g_id IN(SELECT g_id FROM goods_hot);
Query OK, 2 rows affected (0.03 sec)
```

图3-42 删除数据

课 后 习 题

一、选择题

1. 在 SELECT 语句中,用于去除重复行的关键字是(　　)。

 A. LIMIT　　　　B. DISTINCT　　　C. REGEXP　　　D. HAVING

2. 模糊查询使用的关键字是(　　)。

 A. AND　　　　B. NOT　　　　　C. OR　　　　　D. LIKE

3. 在 SELECT 语句中,通配符(　　)表示选择表中所有的列。

 A. ％　　　　　B. ♯　　　　　　C. _　　　　　　D. *

4. 下列聚合函数中用于统计最大值的是(　　)。

 A. SUM(列名)　 B. MIN(列名)　　C. MAX(列名)　　D. COUNT(列名)

5. 限制返回行的数目的关键字是(　　)。

 A. LIMIT　　　　B. JOIN　　　　　C. ALL　　　　　D. COUNT

6. "UPDATE student SET s_name='张丽' WHERE s_id=1;"这一代码执行的操作是(　　)。

 A. 添加姓名叫"张丽"的记录　　　　B. 删除姓名叫"张丽"的记录

 C. 返回姓名叫"张丽"的记录　　　　D. 更新"s_id"值为"1"的姓名为张丽

7. 下列语句不是表数据的基本操作语句的是(　　)。

 A. CREATE 语句　　　　　　　　　B. INSERT 语句

 C. DELETE 语句　　　　　　　　　D. UPDATE 语句

8. 在 SELECT 语句中,可以使用(　　)子句,将结果集的数据行根据列的值进行分组,以便汇总表中内容,实现对每个组的聚合计算。

 A. LIMIT　　　　B. WHERE　　　　C. OEDER BY　　D. GROUP BY

二、简答题

1. WHERE 和 HAVING 都是对查询结果进行筛选,二者有什么区别?

2. 涉及多张数据表时,可以使用连接查询或子查询,二者是否可以互相替代?请说明理由。

项目实训

1. 实训任务

查询"shopping"数据库中的表中数据。

2. 实训目的

(1) 会使用 SELECT 语句查询单表或多表数据。

(2) 会使用 ORDER BY 语句进行数据排序。

(3) 会使用聚合函数进行数据计算。

(4) 会使用 GROUP BY 语句进行数据分组统计。

3. 实训内容

(1) 查询"goods"表中所有信息。

(2) 查询"goods"表中的商品类别、商品名称、商品价格和上架时间。

(3) 查询"users"表中 1995 年 1 月 1 日以后出生的用户的用户名、性别和出生日期,并按照出生日期进行降序排列。

(4) 查询"orders"表中的订单号、订单金额和下单时间。

(5) 查询"goods"表中所有商品的价格总和。

(6) 统计"goods"表中不同类别的商品的库存总量。

(7) 查询用户"张霞"购买的商品的名称、价格和下单时间。

项目四

优化系统数据

默认情况下,数据查询是根据搜索条件进行全表的扫描,将符合条件的记录行显示在结果集中,但随着网上购物系统访问量的增加,全表扫描所需时间会越来越长,数据查询性能会受到影响,甚至会制约应用系统的正常运行。MySQL提供了索引、视图等工具,能有效提高数据查询的效率。本项目主要介绍索引和视图的使用,帮助读者方便快捷地从数据库中查询到所需数据。

学习目标

- 了解索引、视图
- 会创建和管理索引
- 会创建和管理视图

素质目标

由索引和视图的作用,引出利用先进思维方法解决问题的理念,引导学生注重先进理论学习和思维拓展。实操练习能培养学生的动手能力和学习能力,让学生养成主动学习和终身学习的理念。

任务1 索　　引

任务描述

索引是 MySQL 的重要操作对象,主要用于对数据表中的值进行排序,使用索引可以有效提高对数据库中特定数据的查询速度。本任务学习索引的概念和相关操作,包括索引的分类、设计原则,如何创建、查看、维护和删除索引。

4.1.1 索引概述

1. 索引的概念

索引(index)是单独存储在磁盘上,用于快速查找记录的一种数据结构,其作用相当于书的目录。索引是在存储引擎中实现的,在 MySQL 中,存储引擎先在索引中找到对应的值,再根据匹配的索引找到对应的数据。

索引的优点如下:

(1) 加快数据的检索速度,这是创建索引最主要的原因。

(2) 唯一索引可以保证数据表中每行数据的唯一性。

(3) 可以加速表与表之间的连接。

(4) 使用分组和排序进行数据查询时,可以减少分组和排序的时间。

索引的缺点如下:

(1) 创建和维护索引需要消耗时间,随着数据量的增多耗费的时间也会增加。

(2) 索引本身也需要占用磁盘空间。

(3) 对表执行增加、删除、修改等操作时,索引也需要动态维护,这样会降低数据的维护速度。

2. 索引的分类

在 MySQL 中,索引可以分为以下六类。

(1) 主键索引:主键索引是由 PRIMARY KEY 定义的唯一性索引,根据主键的唯一性标识每条记录,避免出现主键索引字段的重复或空值。

(2) 聚集索引:InnoDB 存储引擎中的保存顺序与主键索引顺序一致,这类索引称为聚集索引,一张表中只能有一个聚集索引。

(3) 普通索引:普通索引是 MySQL 中的基本索引,允许在定义索引的列中插入空值和重复值,它的唯一任务就是加快对数据的访问速度。

(4) 唯一索引:唯一索引要求索引列的值必须唯一,允许有空值。若索引包含多个字段,则列值组合必须唯一。

(5) 全文索引:全文索引支持值的全文查找,且允许列值重复或为 NULL,可以在 CHAR、VARCHAR、TEXT 类型的列上创建。

（6）空间索引：空间索引是对空间数据类型的字段创建索引，且索引字段必须声明非空。

3. 索引的设计原则

（1）数据量小的表不要使用索引，因为通过索引查询记录可能比直接扫描整张表还要慢。

（2）不要建立过多的索引，索引并非越多越好，过多的索引会占用大量磁盘空间，并且会降低操作的性能。

（3）对于经常执行修改操作的表不要创建过多索引，且索引中的列也应尽可能少。而对于经常执行查询操作的表，应该创建索引。

（4）在条件表达式中经常会用到的不同值较多的列上创建索引，不同值较少的列不要创建索引。

（5）在频繁进行排序或分组的列上创建索引。

4.1.2 创建和查看索引

1. 使用图形化管理工具 Navicat 创建索引

【例 4-1】 在 Navicat 中，为"goods"表中的"g_type"列创建名为"ix_gtype"的普通索引。操作步骤如下：

（1）打开 Navicat，右键单击资源管理器中"goods"表，选择"设计表"选项，打开"goods"表的设计页面，如图 4-1 所示。

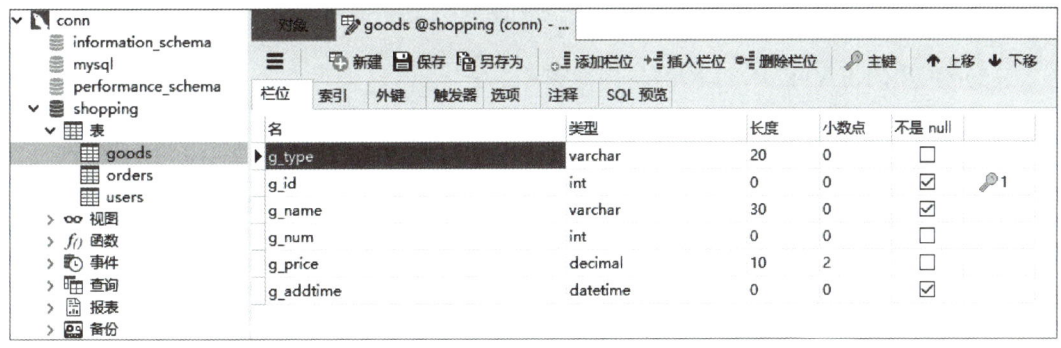

图 4-1　打开设计页面

（2）选择索引选项卡，在"名"中输入"ix_gtype"，"栏位"选择"g_type"，索引类型选择"Normal"，索引方法选择"BTREE"，单击"保存"按钮，如图 4-2 所示。

图 4-2　创建索引

2. 使用 SQL 语句创建索引

（1）使用 CREATE TABLE 语句创建索引

使用 CREATE TABLE 在数据表创建索引，其语法格式为：

CREATE TABLE 表名

（字段定义1,

字段定义2,

……

字段定义n,

索引类型 INDEX/KEY 索引名称（字段名称））

【例 4-2】 创建数据表"new_goods"，为其中的"name"字段创建索引。结果如图 4-3 所示。

```
mysql> USE shopping;
Database changed
mysql> CREATE TABLE new_goods(
    -> id INT(10),
    -> name VARCHAR(30),
    -> price DECIMAL(10,2),
    -> num INT(10),
    -> INDEX n_index(name));
Query OK, 0 rows affected, 2 warnings (0.02 sec)
```

图 4-3　创建索引

（2）使用 CREATE INDEX 语句创建索引

使用 CREATE INDEX 创建索引的语法格式为：

CREATE INDEX 索引名称

ON 表名（字段名称）

【例 4-3】 创建数据表"new_user"，为其中的"name"字段创建唯一索引。结果如图 4-4 所示。

```
mysql> CREATE TABLE new_user(
    -> id INT(10),
    -> name VARCHAR(30),
    -> pwd VARCHAR(30),
    -> gender CHAR(4));
Query OK, 0 rows affected, 1 warning (0.01 sec)

mysql> CREATE UNIQUE INDEX n_unique ON new_user(name);
Query OK, 0 rows affected (0.02 sec)
Records: 0  Duplicates: 0  Warnings: 0
```

图 4-4　创建唯一索引

（3）使用 ALTER TABLE 语句创建索引

使用 ALTER TABLE 语句创建索引的语法格式为：

ALTER TABLE 表名 ADD 索引类型 INDEX/KEY 索引名称（字段名称）

【例 4-4】 使用 ALTER TABLE 语句为"goods"表中的字段"name"创建索引。结果如图 4-5 所示。

```
mysql> ALTER TABLE goods ADD INDEX n_index(g_name);
Query OK, 0 rows affected (0.03 sec)
Records: 0  Duplicates: 0  Warnings: 0
```

图 4-5　为字段"name"创建索引

3. 查看索引信息

在 MySQL 中，通过 SHOW INDEX/KEYS 语句查看索引信息，其语法格式为：

SHOW INDEX/KEYS FROM 表名

【例 4-5】 使用 SHOW INDEX 语句查看"goods"表的索引信息。结果如图 4-6 所示。

```
mysql> SHOW INDEX FROM goods\G
*************************** 1. row ***************************
        Table: goods
   Non_unique: 0
     Key_name: PRIMARY
 Seq_in_index: 1
  Column_name: g_id
    Collation: A
  Cardinality: 8
     Sub_part: NULL
       Packed: NULL
         Null:
   Index_type: BTREE
      Comment:
Index_comment:
      Visible: YES
   Expression: NULL
*************************** 2. row ***************************
        Table: goods
   Non_unique: 1
     Key_name: n_index
 Seq_in_index: 1
  Column_name: g_name
    Collation: A
  Cardinality: 9
     Sub_part: NULL
       Packed: NULL
         Null:
   Index_type: BTREE
      Comment:
Index_comment:
      Visible: NO
   Expression: NULL
2 rows in set (0.00 sec)
```

图 4-6　查看"goods"表的索引信息

4.1.3 维护索引

索引创建后,对数据的操作会引起索引页产生碎片,从而影响数据的查询性能。因此,需要定期对索引进行维护,包括修改索引和删除索引。

1. 修改索引

在 MySQL 中,使用 ALTER TABLE 语句进行索引的修改,其语法格式为:

ALTER TABLE 表名

ALTER INDEX 索引名 VISIBLE/INVISIBLE

【例 4-6】 修改"goods"表中的"n_index"索引为不可见索引(invisible),并查看索引。结果如图 4-7 所示。

图 4-7 修改索引

2. 删除索引

(1) 使用 ALTER TABLE 语句删除索引。使用 ALTER TABLE 语句删除索引的语法格式为:

ALTER TABLE 表名 DROP INDEX 索引名

【例 4-7】 删除"new_goods"表中的索引"n_index"。结果如图 4-8 所示。

图 4-8 删除索引"n_index"

(2) 使用 DROP INDEX 语句删除索引。使用 DROP INDEX 语句删除索引的语法格式为:

DROP INDEX 索引名 ON 表名

【例 4-8】 删除"new_user"表中的索引"n_unique",并使用 SHOW INDEX 语句查看表中索引。结果如图 4-9 所示。

```
mysql> DROP INDEX n_unique ON new_user;
Query OK, 0 rows affected (0.01 sec)
Records: 0  Duplicates: 0  Warnings: 0

mysql> SHOW INDEX FROM new_user\G
Empty set (0.00 sec)
```

图 4-9　删除索引"n_unique"

任务2　视　　图

任务描述

使用 SQL 语句进行数据查询时,结果会在客户端直接输出,但不会被保存。若需要多次使用相同的数据,可以将查询封装成视图,简化查询操作。视图是一种虚拟的表,包含一系列动态生成的数据,用户可以直接使用视图提供的数据。本任务主要学习如何利用视图进行数据查询,从而简化数据操作,提高数据的安全性。

4.2.1　视图概述

1. 视图的定义

视图(view)是从数据库中的一张或多张表中导出的表,其中创建视图时引用的表称为基表。视图和表一样由行和列组成,但它并不存储实际的数据,只是读取基表中的数据。因此,视图中的数据依赖于原始数据表中的数据,若基表中的数据发生变化,视图中的数据也会变化。视图本身是一种虚拟的表,但可以对其进行查询、修改、删除等操作。

2. 视图优点

(1)简单:用户使用视图中的数据,可以不考虑对应表的结构、关联条件等。同时,视图可以把多张表的数据集中在一起,简化了对数据的处理,同时避免了表与表之间的复杂关系。

(2)安全:视图可以作为一种安全机制。通过视图可以进行权限控制,使特定用户只能修改和查询看到的数据;但特定用户既不能看见也不能访问其他关联的表的信息。

(3)数据独立:视图可以使应用程序和数据表在逻辑上保持独立性。创建视图后,当基本表的结构发生变化时,只需要修改视图对应的 SQL 语句,不需要修改应用程序,从而将应用程序与数据表分隔开。

4.2.2 创建和查看视图

1. 使用图形化管理工具 Navicat 创建视图

【例 4-9】 使用 Navicat 创建名称为"view_goods_orders"的视图,列出商品 ID、商品名称、商品价格、用户名、订单号、订单金额和下单日期。操作步骤如下:

(1) 在 Navicat 中打开数据库"shopping",在资源管理器中右键单击"视图"节点,选择"新建视图",如图 4-10 所示。

图 4-10 新建视图

(2) 单击"视图创建工具"。根据本例要求,数据来自三张表,分别是"goods""orders"和"users",将所需的三张表拖到视图主设计区,如图 4-11 所示。

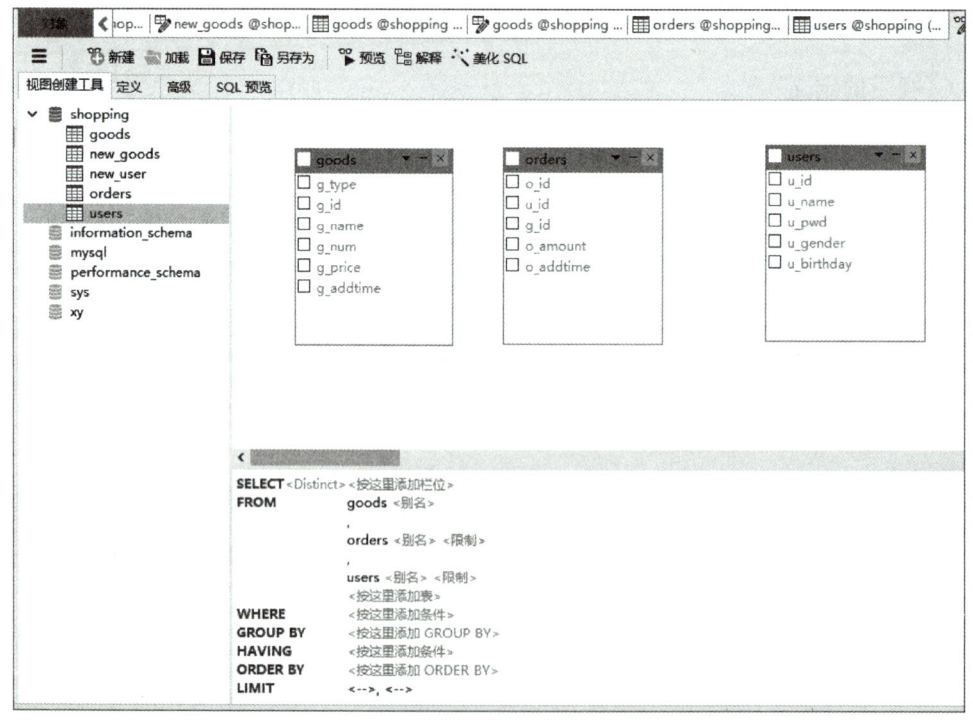

图 4-11 选择数据来源

(3) 根据题目要求,在设计区的各表中勾选要查询的字段,同时建立表与表之间的连接,如图 4-12 所示。

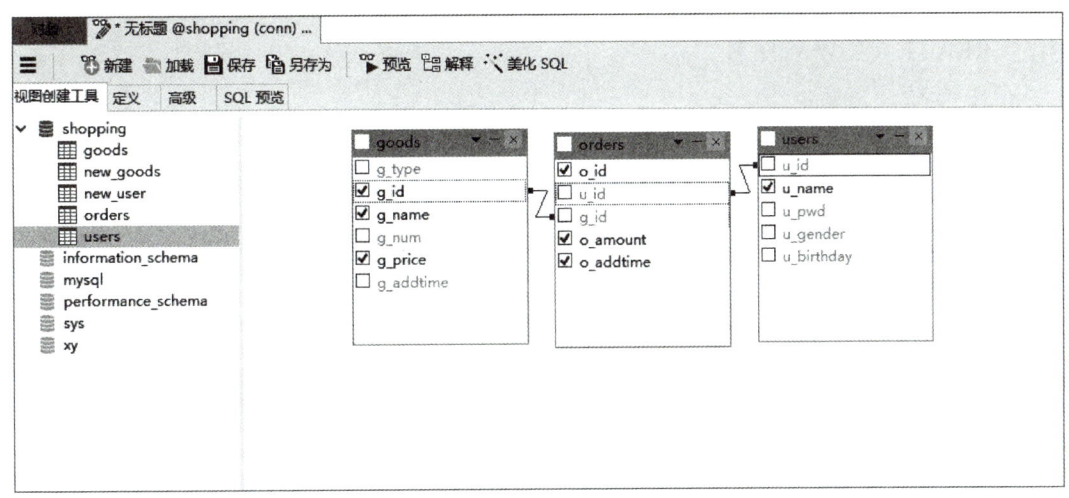

图 4-12　勾选字段

(4) 单击"保存"按钮,输入视图名称"view_goods_orders",单击"确定"按钮,可以在资源管理器中看到视图对象,如图 4-13 所示。

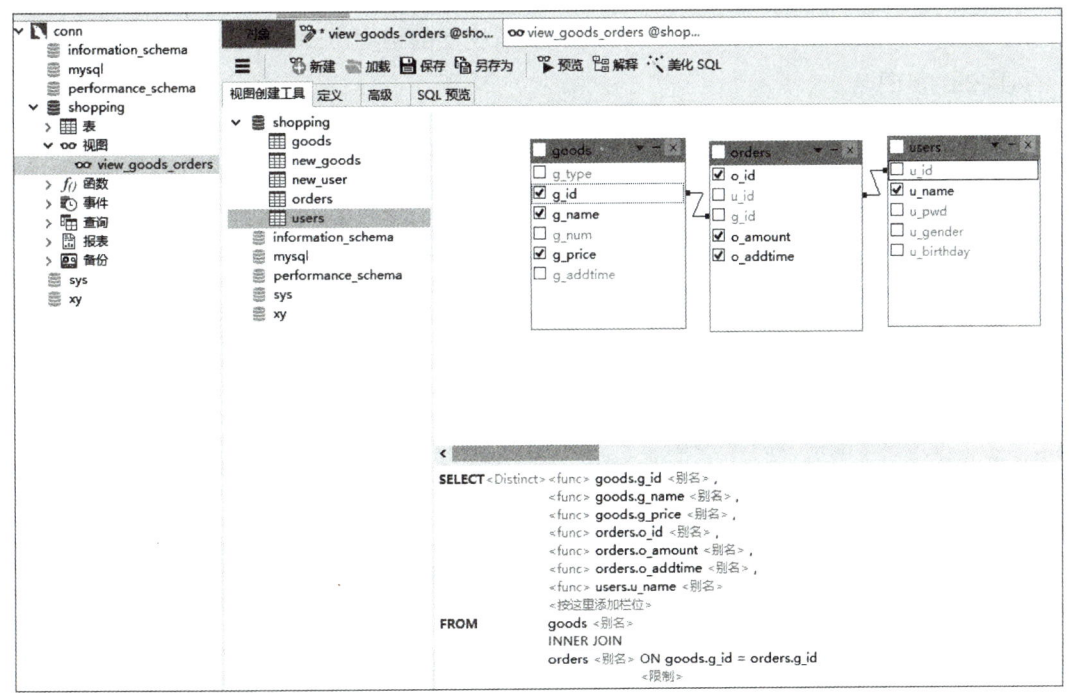

图 4-13　输入视图名称

2. 使用 SQL 语句创建视图

在 MySQL 中使用 CREATE VIEW 语句创建视图，可以在单个表上，也可以在多张表上创建。其语法格式为：

CREATE VIEW 视图名

AS

SELECT 字段名

FROM 表名

【例 4-10】 以"shopping"数据库中"goods"表和"orders"表作为原始数据表，创建名称为"view_goo_ord"的视图，列出商品 ID、商品名称、上架时间、订单 ID，订单金额和下单时间。结果如图 4-14 所示。

```
mysql> CREATE VIEW view_goo_ord(g_id,g_name,g_addtime,o_id,o_amount,o_addtime)
    -> AS
    -> SELECT g.g_id,g.g_name,g.g_addtime,o.o_id,o.o_amount,o.o_addtime
    -> FROM goods g JOIN orders o
    -> ON g.g_id=o.g_id;
Query OK, 0 rows affected (0.00 sec)
```

图 4-14 使用 SQL 语句创建视图

3. 查看视图

（1）在 MySQL 中使用 DESCRIBE 语句查看视图的结构，其语法格式为：

DESC 视图名

【例 4-11】 使用 DESC 语句查看视图"view_goo_ord"的结构。结果如图 4-15 所示。

```
mysql> DESC view_goo_ord;
+-----------+--------------+------+-----+---------+-------+
| Field     | Type         | Null | Key | Default | Extra |
+-----------+--------------+------+-----+---------+-------+
| g_id      | int          | NO   |     | NULL    |       |
| g_name    | varchar(30)  | NO   |     | NULL    |       |
| g_addtime | datetime     | NO   |     | NULL    |       |
| o_id      | int          | NO   |     | NULL    |       |
| o_amount  | decimal(20,2)| YES  |     | NULL    |       |
| o_addtime | datetime     | YES  |     | NULL    |       |
+-----------+--------------+------+-----+---------+-------+
6 rows in set (0.00 sec)
```

图 4-15 查看视图的结构

（2）使用 SHOW CREATE VIEW 语句可以查看视图的定义语句，其语法格式为：

SHOW CREATE VIEW 视图名

【例 4-12】 使用 SHOW CREATE VIEW 语句查看视图"view_goo_ord"的定义语句。结果如图 4-16 所示。

```
mysql> SHOW CREATE VIEW view_goo_ord\G;
*************************** 1. row ***************************
                View: view_goo_ord
         Create View: CREATE ALGORITHM=UNDEFINED DEFINER=`root`@`localhost` SQL SECURITY DEFINER VIEW `view_goo_ord` (`g_id`,`g_name`,`g_addtime`,`o_id`,`o_a
mount`,`o_addtime`) AS select `g`.`g_id` AS `g_id`,`g`.`g_name` AS `g_name`,`g`.`g_addtime` AS `g_addtime`,`o`.`o_id` AS `o_id`,`o`.`o_amount` AS `o_amount`,
`o`.`o_addtime` AS `o_addtime` from (`goods` `g` join `orders` `o` on((`g`.`g_id` = `o`.`g_id`)))
character_set_client: gbk
collation_connection: gbk_chinese_ci
1 row in set (0.00 sec)
```

注：以"\G"结尾，其作用是将查询结果按列显示。

图 4-16　查看视图的定义语句

4.2.3　维护视图

当用户需求发生变化时，可以通过图形化管理工具 Navicat 和 SQL 语句修改和删除视图，利用 Navicat 修改和删除视图参照创建视图操作步骤，这里不再赘述。

1. 修改视图

（1）使用 CREATE OR REPLACE VIEW 语句修改视图。

使用 CREATE OR REPLACE VIEW 语句修改视图的语法格式为：

CREATE OR REPLACE VIEW 视图名（字段名）

AS

SELECT 字段名

FROM 表名

【例 4-13】　修改视图"view_goo_ord"，在原有查询基础上删除商品上架时间，添加商品价格，执行并查询结果，如图 4-17 所示。

```
mysql> CREATE OR REPLACE VIEW view_goo_ord(g_id,g_name,g_price,o_id,o_amount,o_addtime)
    -> AS
    -> SELECT g.g_id,g.g_name,g.g_price,o.o_id,o.o_amount,o.o_addtime
    -> FROM goods g JOIN orders o
    -> ON g.g_id=o.g_id;
Query OK, 0 rows affected (0.01 sec)

mysql> DESC view_goo_ord;
+-----------+---------------+------+-----+---------+-------+
| Field     | Type          | Null | Key | Default | Extra |
+-----------+---------------+------+-----+---------+-------+
| g_id      | int           | NO   |     | NULL    |       |
| g_name    | varchar(30)   | NO   |     | NULL    |       |
| g_price   | decimal(10,2) | YES  |     | NULL    |       |
| o_id      | int           | NO   |     | NULL    |       |
| o_amount  | decimal(20,2) | YES  |     | NULL    |       |
| o_addtime | datetime      | YES  |     | NULL    |       |
+-----------+---------------+------+-----+---------+-------+
6 rows in set (0.00 sec)
```

图 4-17　修改视图"view_goo_ord"

（2）使用 ALTER 语句修改视图。

使用 ALTER 语句修改视图的语法格式为：

ALTER VIEW 视图名（字段名）

AS

SELECT 字段名

FROM 表名

【例 4-14】 修改视图"view_goods",在原有查询基础上删除商品数量,添加商品上架时间,执行并查询的结果如图 4-18 所示。

```
mysql> ALTER VIEW view_goods(g_id,g_name,g_type,g_price,g_addtime)
    -> AS
    -> SELECT g_id,g_name,g_type,g_price,g_addtime
    -> FROM goods;
Query OK, 0 rows affected (0.00 sec)

mysql> DESC view_goods;
+-----------+---------------+------+-----+---------+-------+
| Field     | Type          | Null | Key | Default | Extra |
+-----------+---------------+------+-----+---------+-------+
| g_id      | int           | NO   |     | NULL    |       |
| g_name    | varchar(30)   | NO   |     | NULL    |       |
| g_type    | varchar(20)   | YES  |     | NULL    |       |
| g_price   | decimal(10,2) | YES  |     | NULL    |       |
| g_addtime | datetime      | NO   |     | NULL    |       |
+-----------+---------------+------+-----+---------+-------+
5 rows in set (0.00 sec)
```

图 4-18 修改视图"view_goods"

2. 删除视图

在 MySQL 中使用 DROP VIEW 语句删除视图,其语法格式为:

DROP VIEW [IF EXISTS]视图 1 名称,视图 2 名称,……,视图 n 名称

【例 4-15】 删除视图"view_goods",执行并查看结果,如图 4-19 所示。

```
mysql> DROP VIEW view_goods;
Query OK, 0 rows affected (0.01 sec)

mysql> DESC view_goods;
ERROR 1146 (42S02): Table 'shopping.view_goods' doesn't exist
mysql>
```

图 4-19 删除视图"view_goods"

4.2.4 通过视图操作数据

1. 向视图中插入数据

在 MySQL 中可以使用 INSERT 语句向视图中插入数据,其语法格式为:

INSERT INTO 视图名(字段名)

VALUES(字段 1 值,字段 2 值,字段 3 值……)

【例 4-16】 向视图"view_users"中插入数据。结果如图 4-20 所示。

```
mysql> INSERT INTO view_users
    -> VALUES(8,'王明','wm123','男','1980-03-03 10:08:40');
Query OK, 1 row affected (0.00 sec)
```

图 4-20 向视图"view_users"中插入数据

查看插入后的视图。结果如图 4-21 所示。

```
mysql> SELECT *
    -> FROM view_users;
+------+--------+-------------+----------+---------------------+
| u_id | u_name | u_pwd       | u_gender | u_birthday          |
+------+--------+-------------+----------+---------------------+
|    1 | 李向阳 | lixiangyang | 男       | 1990-07-22 11:53:51 |
|    2 | 张霞   | zx123       | 女       | 1993-08-30 08:54:30 |
|    3 | 刘晓宇 | lix567      | 男       | 2000-02-11 15:55:24 |
|    4 | 周丽   | zhouli123   | 女       | 1998-11-01 02:30:07 |
|    5 | 李雅萍 | liyp456     | 女       | 1991-10-07 11:51:37 |
|    6 | 张辉   | zhanghui000 | 男       | 1987-04-11 08:52:34 |
|    8 | 王明   | wm123       | 男       | 1980-03-03 10:08:40 |
+------+--------+-------------+----------+---------------------+
7 rows in set (0.00 sec)
```

图 4-21　查看插入后的视图

2. 修改视图中的数据

在 MySQL 中可以使用 UPDATE 语句更新视图中的数据，其语法格式为：

UPDATE 视图名

SET 字段名 1＝值 1，字段名 2＝值 2，……，字段名 n＝值 n

WHERE 条件表达式

【例 4-17】　修改视图"view_users"中的数据，将用户名为"李向阳"的密码修改为"lxy723"，执行并查看结果，如图 4-22 所示。

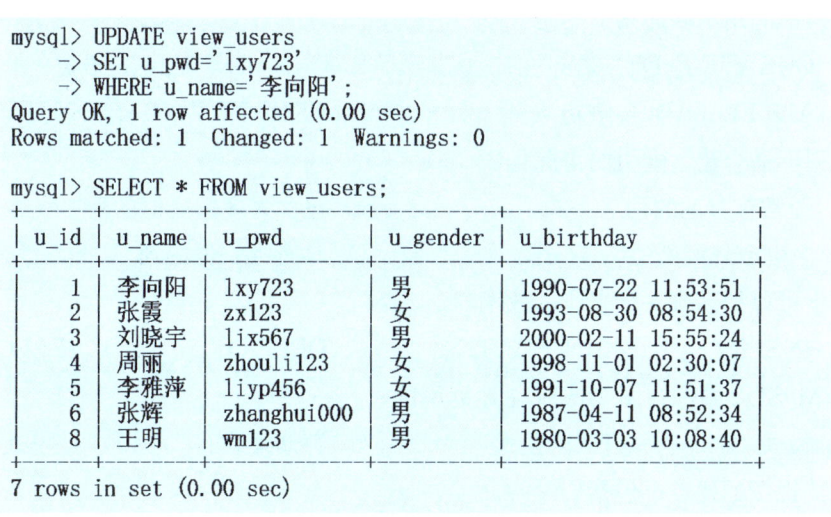

图 4-22　修改视图中的数据

3. 删除视图中的数据

在 MySQL 中可以使用 DELETE 语句删除视图中的数据，其语法格式为：

DELETE FROM 视图名

WHERE 条件表达式

【例 4-18】 删除视图"view_users"中用户名为"张辉"的数据。执行结果如图 4-23 所示。

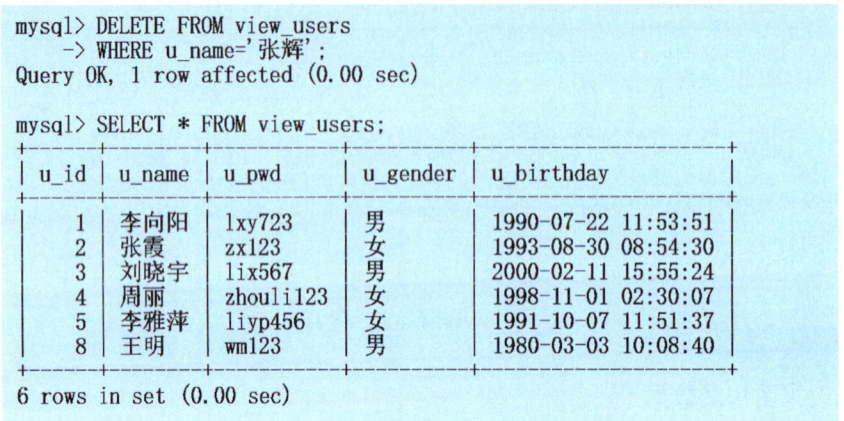

图 4-23 删除视图中的数据

课后习题

一、选择题

1. 下列不能用于创建索引的是（　　）。
   A. CREATE INDEX 语句　　　　　　B. CREATE TABLE 语句
   C. ALTER TABLE 语句　　　　　　　D. CREATE DATABASE 语句
2. 下列不适合建立索引的情况是（　　）
   A. 经常被查询的列　　　　　　　　B. 包含太多重复值的列
   C. 主键或外键列　　　　　　　　　D. 具有唯一值的列
3. 索引可以提高哪一种操作的效率？（　　）
   A. INSERT　　　B. UPDATE　　　C. DELETE　　　D. SELECT
4. 在 MySQL 中的视图 VIEW 是数据库的（　　）。
   A. 模式　　　　B. 内模式　　　　C. 存储模式　　　D. 外模式
5. 创建视图的优点不包括（　　）。
   A. 简化用户对数据的查询与处理　　B. 提高数据查询效率
   C. 方便对数据进行用户权限管理　　D. 提高逻辑数据独立性
6. 以下不可对视图执行的操作有（　　）。
   A. SELECT　　　　　　　　　　　　B. INSERT
   C. CREATE INDEX　　　　　　　　　D. DELETE

二、简答题

1. 简述索引的设计原则。
2. 举例说明视图是如何保障数据安全性的。

项 目 实 训

1. 实训任务

（1）创建索引、查看索引和维护索引。

（2）创建视图、管理视图和维护视图。

（3）通过视图操作数据。

2. 实训目的

（1）能分别使用 Navicat 和 SQL 语句创建索引。

（2）能使用 SQL 语句查看索引。

（3）能分别使用 Navicat 和 SQL 语句创建视图。

（4）能使用 SQL 语句维护视图。

（5）能使用 SQL 语句在视图中操作数据。

3. 实训内容

（1）在"shopping"数据库中利用图形化管理工具 Navicat 在表"goods""users"和"orders"上创建普通索引、唯一性索引和复合索引。

（2）使用 SQL 语句在表"goods""users"和"orders"上创建普通索引、唯一性索引和复合索引。

（3）使用 SQL 语句查看(2)中创建的索引。

（4）使用 SQL 语句修改和删除(2)中创建的索引。

（5）在"shopping"数据库中利用图形化管理工具 Navicat 在表"goods""users"和"orders"上创建视图。

（6）使用 SQL 语句在表"goods""users"和"orders"上创建视图。

（7）使用 SQL 语句查看(6)中创建的视图。

（8）使用 SQL 语句修改和删除(6)中创建的视图。

（9）使用 SQL 语句在视图中添加、修改和删除数据。

项目五

数据库编程

数据库编程有效解决了程序设计中的复杂逻辑问题,提高了数据访问效率。本项目介绍数据库编程中存储过程、存储函数和触发器等数据操作对象及其在数据库应用系统开发中的使用方法,实现数据库的程序模块化设计。

学习目标

- 了解存储过程、存储函数、触发器
- 会创建和调用存储过程
- 会创建和调用存储函数
- 会创建和调用触发器

素质目标

对存储过程、存储函数和触发器的运用能引导学生化繁为简,把复杂的问题分解,简单化,启发学生在学习和生活中做好规划,稳步前进。

项目五 数据库编程

任务 1 存储过程

任务描述

存储过程是数据库的重要语句集合,能够将特定的语句块进行封装,完成特定操作和任务,从而提高程序的复用性。本任务主要从存储过程的优点着手,学习创建、调用存储过程的方法。

5.1.1 存储过程概述

1. 存储过程的概念

存储过程是一组经过编译并保存在数据库中的 SQL 语句集合,可以随时被调用。存储过程采用预编译方式,当执行一次后,其执行计划就保留在缓存中,再次调用时可以直接使用已编译好的语句,因此执行效率较高。

2. 存储过程的优点

存储过程可以完成特定功能的封装,减少客户端与服务器端的数据传输,提高数据访问效率。其有以下四个优点:

(1) 存储过程执行一次后,其执行计划就保留在缓存中,再次调用只需要从高速缓冲存储器中调用,提高了系统的性能。

(2) 存储过程的数据返回可以通过 SELECT 语句和输出函数实现。

(3) 存储过程的应用灵活,且可以嵌套在触发器或事件中。

(4) 数据库的管理员可以对存储过程进行单独的权限控制,避免非授权用户对数据的访问,从而保证数据的安全性。

5.1.2 创建存储过程

1. 使用 Navicat 图形化管理工具创建存储过程

【例 5-1】 使用 Navicat 图形化管理工具创建存储过程,其操作步骤如下:

(1) 打开 Navicat 图形化管理工具中的数据库"shopping",在资源管理器中右键单击"函数"选项,选择"新建"函数,如图 5-1 所示。

(2) 选择存储过程,单击左下角"+"按钮,在编辑区域添加所需的参数,单击"完成"按钮,如图 5-2 所示。

(3) 在 BEGIN…END 语句中编辑需要执行的 SQL 语句(图 5-3),单击"保存"按钮,输入存储过程名称即可。

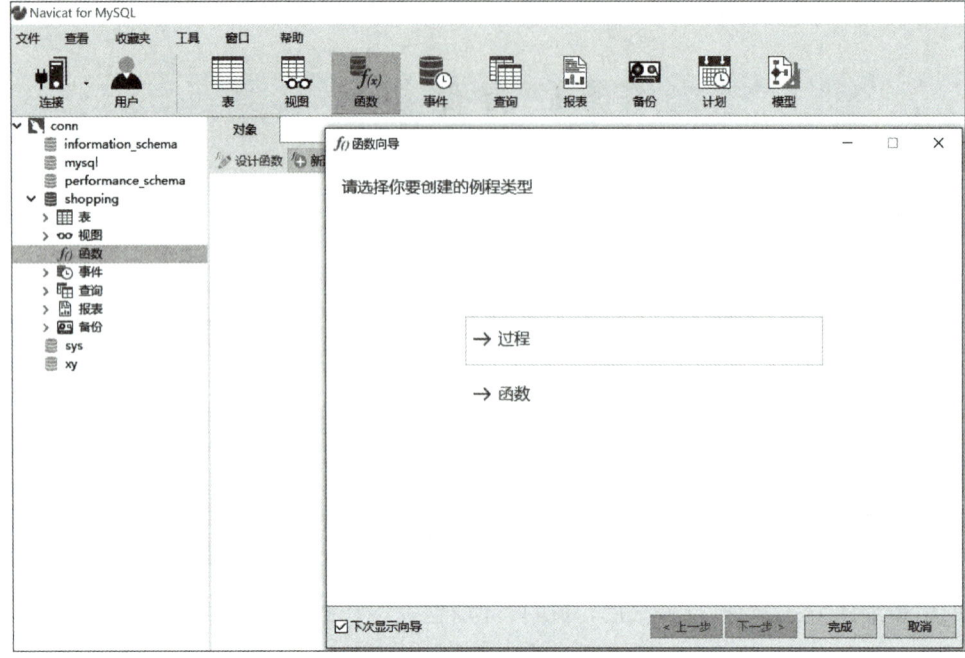

图 5-1 新建过程

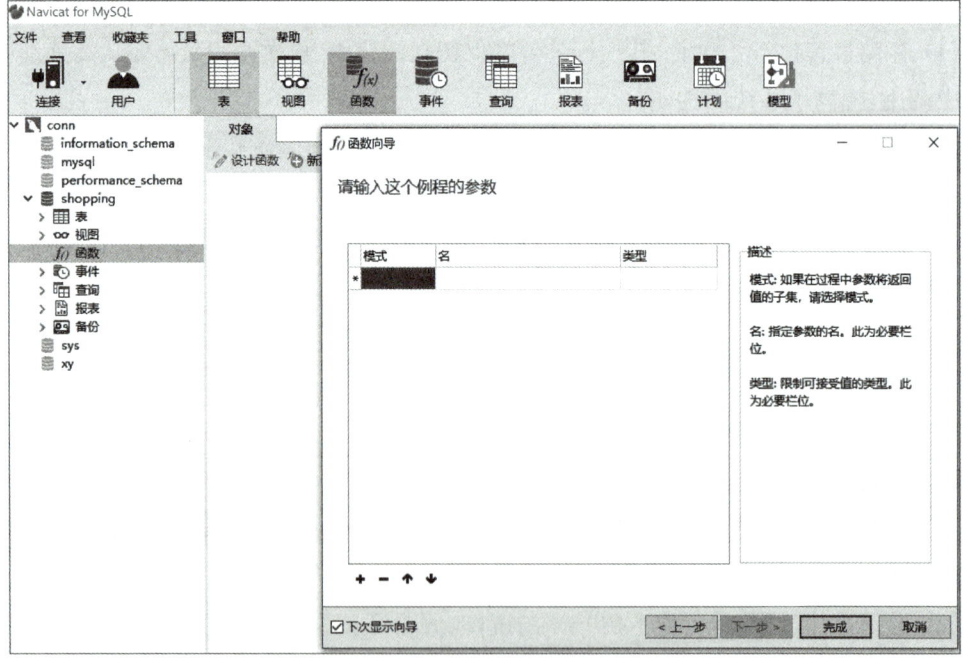

图 5-2 添加参数

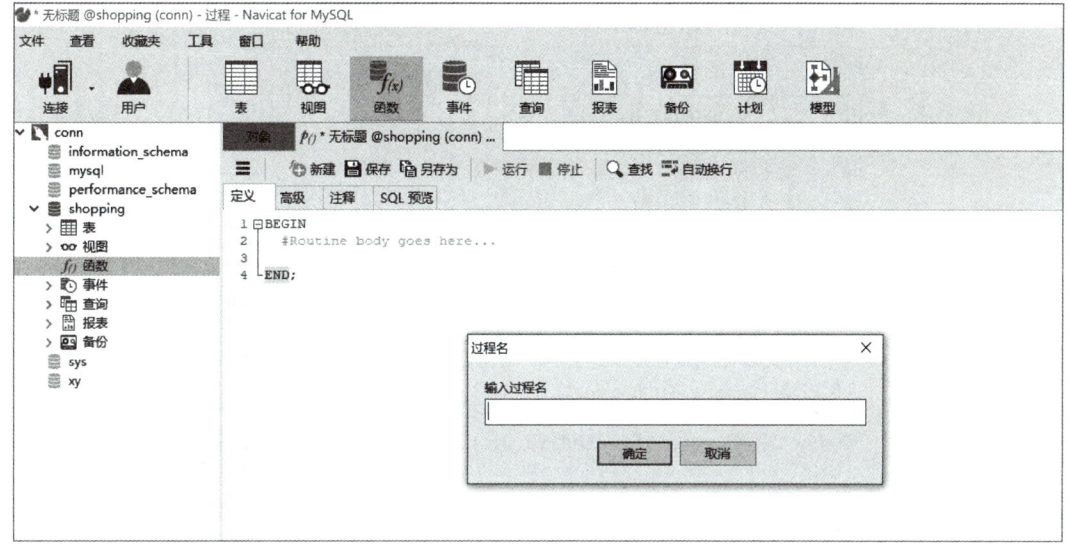

图 5-3　编辑 SQL 语句

2. 使用 SQL 语句创建存储过程

在 MySQL 中，创建存储过程的基本语法格式为：

CREATE PROCEDURE 存储过程名（参数）

程序体

【例 5-2】　创建存储过程"pro_getgoods"，查询"goods"表中的商品 ID、商品名称、商品数量和商品价格。结果如图 5-4 所示。

```
mysql> DELIMITER $$
mysql> CREATE PROCEDURE pro_getgoods()
    -> READS SQL DATA
    -> BEGIN
    -> SELECT g_id,g_name,g_num  FROM goods;
    -> END $$
Query OK, 0 rows affected (0.01 sec)
```

图 5-4　创建存储过程

5.1.3　调用存储过程

在 MySQL 中使用 CALL 语句来调用存储过程，调用后，数据库将执行存储过程中的语句，并将执行结果返回并输出。其语法格式为：

CALL 存储过程名（参数）

【例 5-3】　调用名称为"pro_getgoods"的存储过程，并列出相应商品的商品信息。结果如图 5-5 所示。

```
mysql> CALL pro_getgoods;
+------+----------------------------------+-------+
| g_id | g_name                           | g_num |
+------+----------------------------------+-------+
|    1 | 华为P50手机                       |    10 |
|    2 | 四大名著原著版                     |    20 |
|    3 | 陕西洛川红富士苹果礼盒装            |    40 |
|    4 | iPhone手机                        |  NULL |
|    5 | 托尔斯泰三部曲                     |    40 |
|    6 | 高尔基三部曲                       |    70 |
|    7 | 三只松鼠干果礼盒装                  |  NULL |
|    8 | 西湖龙井礼盒装                     |  NULL |
|    9 | 小米手机                          |  NULL |
|   10 | 手机充电器                         |    10 |
+------+----------------------------------+-------+
10 rows in set (0.00 sec)

Query OK, 0 rows affected (0.01 sec)
```

图 5-5 调用存储过程

存储过程的查看、修改和删除与表、索引等操作类似，这里不再赘述。

任务 2　存 储 函 数

🔹任务描述

在实际应用中，为了让应用程序专注业务处理，数据库定义了存储函数来封装数据处理逻辑，以有效实现数据库中的数据访问。本任务主要学习存储函数的创建与调用过程。

5.2.1　存储函数概述

存储函数和存储过程类似，也是在数据库中定义的能完成特定功能的语句集合，用户可以通过自定义存储函数来完成特定功能，进而避免重复编写相同的 SQL 语句，减少客户端和服务器端的数据传输。

5.2.2　创建存储函数

1. 使用 Navicat 图形化管理工具创建存储函数

【例 5-4】 使用 Navicat 图形化管理工具创建存储函数，其操作步骤如下：

（1）打开 Navicat 图形化管理工具中的数据库"shopping"，在资源管理器中右键单击"函数"选项，选择"新建"函数，如图 5-6 所示。

（2）选择存储函数，单击左下角"+"按钮，在编辑区域添加所需的参数，单击"完成"按钮，如图 5-7 所示。

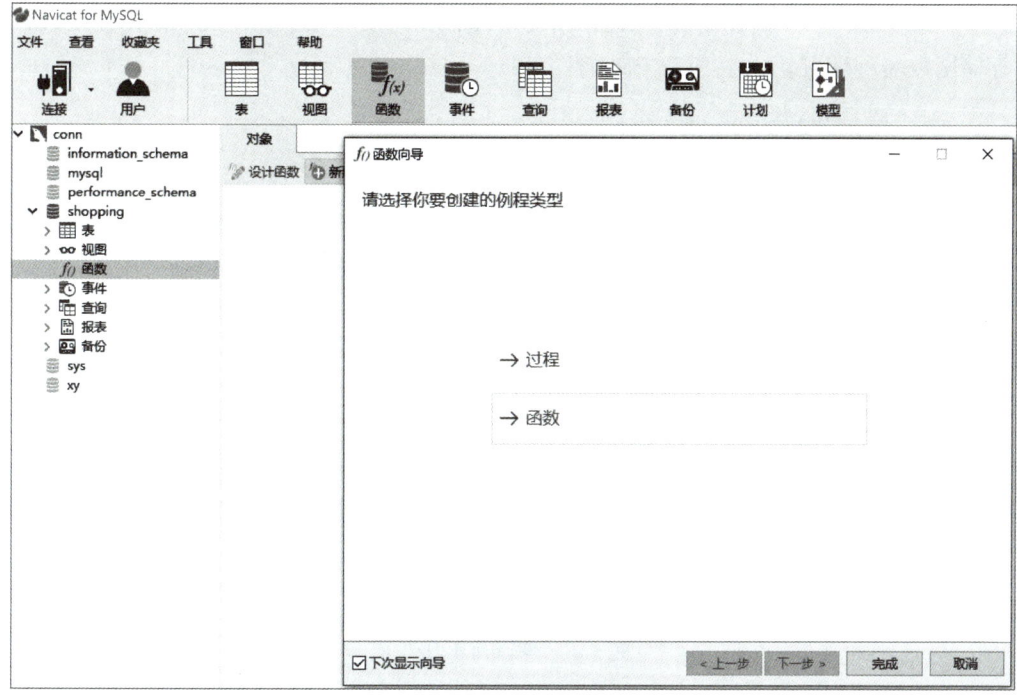

图 5-6　新建存储函数

图 5-7　添加参数

(3) 在 BEGIN…END 语句中编辑需要执行的 SQL 语句(图 5-8)，单击"保存"按钮，输入存储函数名称即可。

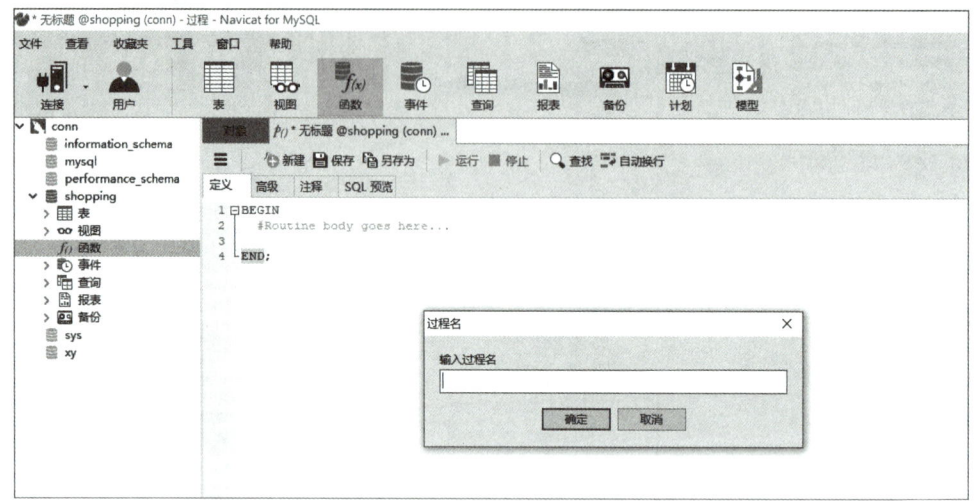

图 5-8　编辑 SQL 语句

2. 使用 SQL 语句创建存储函数

在 MySQL 中，创建存储函数的基本语法格式为：

CREATE FUNCTION 函数名(参数)

RETURNS 数据类型

函数体

【例 5-5】 创建存储函数"func_goods"，返回"goods"表中商品的数目。结果如图 5-9 所示。

```
mysql> CREATE FUNCTION func_goods ()
    -> RETURNS integer
    -> DETERMINISTIC
    -> RETURN (SELECT COUNT(*) FROM goods);
Query OK, 0 rows affected (0.00 sec)
```

图 5-9　创建存储函数"func_goods"

【例 5-6】 创建存储函数"func_getname"，查询商品名称。结果如图 5-10 所示。

```
mysql> CREATE FUNCTION func_getname()
    -> RETURNS varchar(30)
    -> DETERMINISTIC
    -> RETURN (SELECT g_name FROM goods);
Query OK, 0 rows affected (0.00 sec)
```

图 5-10　创建存储函数"func_getname"

5.2.3 调用存储函数

在 MySQL 中,存储函数的使用方法与内置函数一致,区别是存储函数由用户自定义,而内置函数由 MySQL 开发者定义。调用存储函数的语法格式为:

SELECT 函数名(参数)

【例 5-7】 调用名称为"func_goods"的存储函数,返回"goods"表中商品的数目。结果如图 5-11 所示。

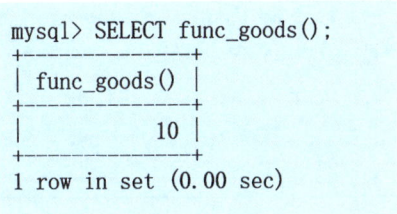

图 5-11 调用存储函数"func_goods"

存储函数的查看、修改和删除与表、索引等操作类似,这里不再赘述。

任务3 触 发 器

任务描述

触发器是数据库的操作对象之一,是一段程序代码。当数据表中出现特定事件时,就会激发该对象。为了保证数据完整性,设计人员可以通过触发器实现复杂的业务逻辑。本任务主要学习创建触发器和管理触发器的方法。

5.3.1 触发器概述

在 MySQL 中,触发器是存储在系统内的一段程序代码,也可以看成是一个特殊的存储过程,用来对表实施约束,以保持数据的一致性。触发器无须人工调用,当程序满足定义条件时会被自动调用。MySQL 中激活触发器的操作包括 INSERT、UPDATE 和 DELETE。

触发器定义在表上,MySQL 提供新旧两张逻辑表,旧表用来存放更新前的记录,新表用来存放更新后的记录。当触发器执行完成之后,两张表会被自动删除。

5.3.2 创建触发器

1. 使用 Navicat 图形化管理工具创建触发器

【例 5-8】 使用 Navicat 图形化管理工具创建触发器,其操作步骤如下:

(1) 在图形化管理工具 Navicat 中选择数据库"shopping"中要创建触发器的表"orders",右键单击"设计表"按钮,选择"触发器"选项卡,如图 5-12 所示。

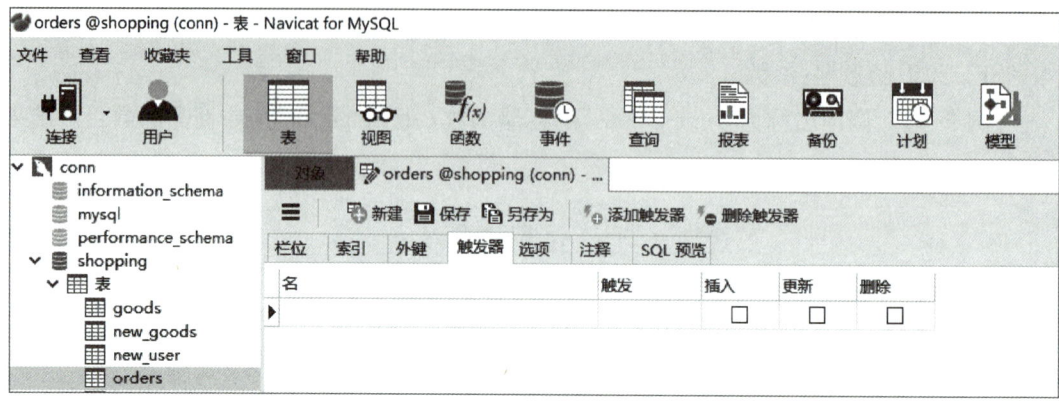

图 5-12　创建触发器

(2) 选择"添加触发器",输入触发器名称和基本特性,在下方定义区域编辑出触发器激活后要执行的 SQL 语句,单击"保存"按钮,如图 5-13 所示。

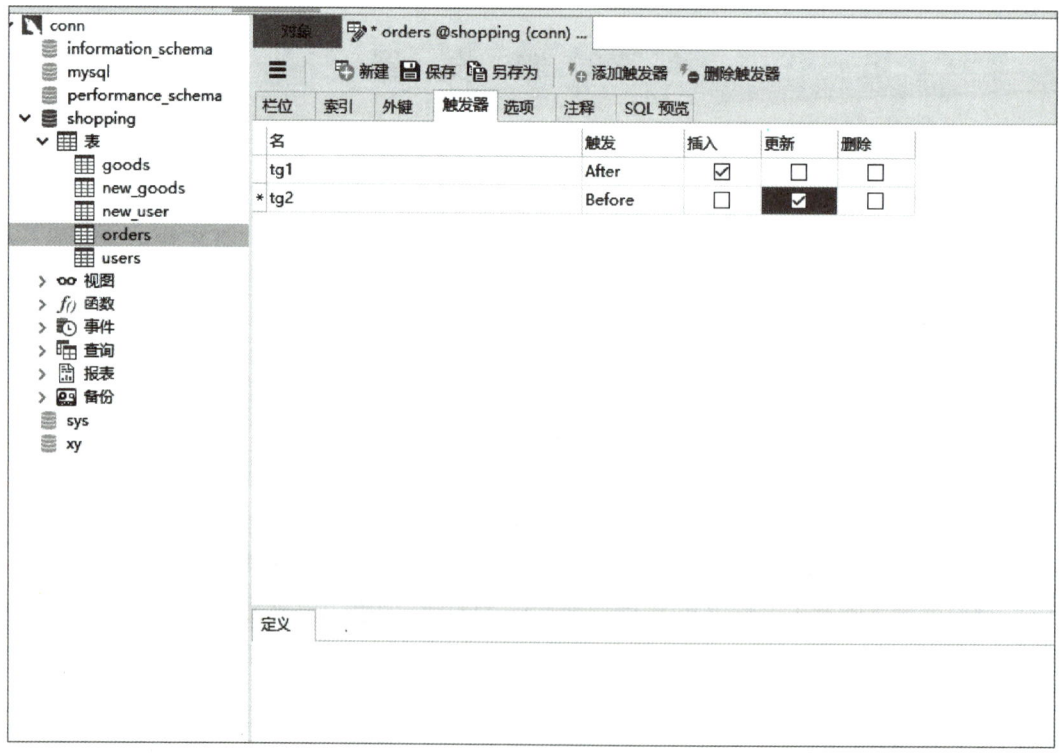

图 5-13　添加触发器

2. 使用 SQL 语句创建触发器

使用 SQL 语句创建触发器的语法格式为：

CREATE TRIGGER 触发器名称

触发时间 触发事件

ON 表名

FOR EACH ROW

程序体

注：这里的触发时间取值为 BEFORE 或者 AFTER，BEFORE 是在检查约束之前触发，AFTER 是在检查约束之后触发；触发事件包含 INSERT、UPDATE、DELETE。

【例 5-9】 为订单表"orders"创建 AFTER 触发器，在向订单表中插入记录的同时，自动更新商品信息表中的数量，并验证其应用。操作结果如图 5-14 所示。

```
mysql> CREATE TRIGGER tg1
    -> AFTER INSERT ON orders
    -> FOR EACH ROW
    -> UPDATE goods SET g_num=g_num-1 WHERE g_id=1;
Query OK, 0 rows affected (0.01 sec)
```

图 5-14　使用 SQL 语句创建 AFTER 触发器

当向订单表"orders"中插入记录后，商品信息表"goods"的数量自动减 1。结果如图 5-15 所示。

```
mysql> INSERT INTO orders(o_id,u_id,g_id,o_num) VALUES(8,1,1,1);
Query OK, 1 row affected (0.00 sec)

mysql> SELECT * FROM goods;
+----------+------+------------------------------+-------+---------+---------------------+
| g_type   | g_id | g_name                       | g_num | g_price | g_addtime           |
+----------+------+------------------------------+-------+---------+---------------------+
| 电子产品 |    1 | 华为P50手机                  |     8 | 4258.00 | 2021-05-01 11:33:40 |
| 书籍     |    2 | 四大名著原著版               |    20 |  450.00 | 2021-08-10 17:36:07 |
| 礼品     |    3 | 陕西洛川红富士苹果礼盒装     |    40 |   56.20 | 2022-06-16 12:36:54 |
| 电子产品 |    4 | iPhone手机                   |  NULL | 8000.00 | 2022-01-01 15:32:34 |
| 书籍     |    5 | 托尔斯泰三部曲               |    40 |   92.00 | 2021-07-12 12:53:26 |
| 书籍     |    6 | 高尔基三部曲                 |    70 |   44.60 | 2021-10-20 13:27:25 |
| 礼品     |    7 | 三只松鼠干果礼盒装           |  NULL |  380.00 | 2022-04-26 14:25:22 |
| 礼品     |    8 | 西湖龙井礼盒装               |  NULL |  160.00 | 2022-05-30 15:07:08 |
| 电子产品 |    9 | 小米手机                     |  NULL | 2000.00 | 2022-05-09 10:55:21 |
| 电子产品 |   10 | 手机充电器                   |    10 |  100.00 | 2022-11-16 13:58:54 |
+----------+------+------------------------------+-------+---------+---------------------+
10 rows in set (0.00 sec)
```

图 5-15　触发结果

5.3.3　查看触发器

在 MySQL 中使用 SHOW TRIGGERS 语句来查看触发器，其语法格式为：

SHOW TRIGGERS LIKE 匹配模式|WHERE 条件表达式\G

【例 5-10】 查看"orders"表上的触发器。结果如图 5-16 所示。

```
mysql> SHOW TRIGGERS LIKE 'orders%'\G;
*********************** 1. row ***********************
             Trigger: tg1
               Event: INSERT
               Table: orders
           Statement: UPDATE goods SET g_num=g_num-1 WHERE g_id=1
              Timing: AFTER
             Created: 2022-11-21 15:06:43.14
            sql_mode: STRICT_TRANS_TABLES,NO_ENGINE_SUBSTITUTION
             Definer: root@localhost
character_set_client: gbk
collation_connection: gbk_chinese_ci
  Database Collation: utf8mb4_0900_ai_ci
1 row in set (0.00 sec)
```

图 5-16　查看触发器

5.3.4　删除触发器

在 MySQL 中使用 DROP TRIGGERS 语句来删除触发器,其语法格式为:
DROP TRIGGERS 触发器名称

【例 5-11】 删除"orders"表中名称为"tg1"的触发器。结果如图 5-17 所示。

```
mysql> DROP TRIGGER tg1;
Query OK, 0 rows affected (0.01 sec)
```

图 5-17　删除触发器

课后习题

一、选择题

1. 存储函数中选择语句使用(　　)。
   A. IF　　　　　B. WHILE　　　　C. SELECT　　　　D. SWITCH
2. MySQL 使用(　　)来调用存储过程。
   A. EXEC　　　　B. CALL　　　　C. EXECUTE　　　　D. CREATE
3. 激活触发器的事件包括 INSERT、UPDATE 和(　　)事件。
   A. CREATE　　　B. ALTER　　　C. DROP　　　　D. DELETE
4. 下列关于删除存储过程的语句正确的是(　　)。
   A. DROP PROC proc1　　　　　　B. DELETE PROC proc1
   C. DROP PROCEDURE proc1　　　D. DELETE PROCEDURE proc1

5. 下列操作中,不可能触发对应关系表上触发器的操作是(　　)。

　　A. SELECT　　　B. INSERT　　　C. UPDATE　　　D. DELETE

二、思考题

1. 使用存储过程的优缺点有哪些?

2. 使用触发器时会触发一些数据操作,因此需要特别小心,使用触发器需要注意哪些问题?

项目实训

1. 实训任务

(1) 创建和调用存储过程。

(2) 创建和调用存储函数。

(3) 创建和管理触发器。

2. 实训目的

(1) 能分别使用 Navicat 和 SQL 语句创建存储过程。

(2) 能使用 SQL 语句调用存储过程。

(3) 能分别使用 Navicat 和 SQL 语句创建和调用存储函数。

(4) 能分别使用 Navicat 和 SQL 语句创建触发器。

(5) 能使用 SQL 语句查看和管理触发器。

3. 实训内容

(1) 在"shopping"数据库中利用图形化管理工具 Navicat 和 SQL 语句在"goods"表上创建存储过程。

(2) 使用"SQL"语句调用存储过程。

(3) 使用"SQL"语句修改和删除存储过程。

(4) 在"shopping"数据库中利用图形化管理工具 Navicat 和 SQL 语句在"orders"表上创建存储函数。

(5) 使用 SQL 语句调用存储函数。

(6) 使用 SQL 语句管理存储函数。

(7) 在"shopping"数据库中利用图形化管理工具 Navicat 和 SQL 语句在"goods"表上创建触发器。

(8) 使用 SQL 语句管理触发器。

(9) 使用 SQL 语句修改和删除触发器。

项目六

数据库系统的安全性和可用性

随着信息化、网络化水平的不断提升,越来越多的数据保存在数据库系统中,数据已然成为系统运行的核心,受到人们的广泛关注。MySQL 提供了用户认证、授权、备份、日志等机制,以维护和实现数据的安全,避免用户越权访问数据库,有效防止因不可抗拒的因素导致的数据丢失。本项目主要介绍用户权限、数据库备份和恢复等机制,保障数据的安全性和可用性。

学习目标

- 能在数据库中创建、删除用户
- 能对数据库中的权限进行授予、查看和回收操作
- 会备份和恢复数据
- 会使用日志恢复数据

素质目标

由数据安全性出发,引导学生树立信息安全意识,培养学生的职业道德规范和社会责任感。

任务 1　用户权限管理

> **任务描述**
>
> 随着信息化水平不断提高,信息安全越来越受到人们的关注。MySQL 是多用户数据库,可以为不同用户分配相应的访问数据库对象及数据的权限,提供用户认证、授权等机制来维护数据的安全,有效防范信息泄露、更改和破坏。本任务主要学习 MySQL 中用户及用户权限管理的实现。

6.1.1　用户权限

数据库的安全性指允许合法用户进行其权限范围内的数据库相关操作,防止不合法的使用所造成的数据泄露、更改或破坏。其中,数据库安全性主要涉及用户认证和访问权限两个方面。

MySQL 用户主要包括 root 用户和普通用户。root 用户是超级管理员,拥有创建用户、删除用户、修改用户密码等所有数据库权限,普通用户仅拥有被 root 用户授予的各种权限。

在安装 MySQL 时,会自动安装名为 mysql 的系统数据库(图 6-1),里面包含 user 的账户信息和一些全局级的权限信息,主要分为 6 个类别,分别是账号列、安全连接列、身份验证和密码策略列、资源控制列、权限列和用户特征数据列。

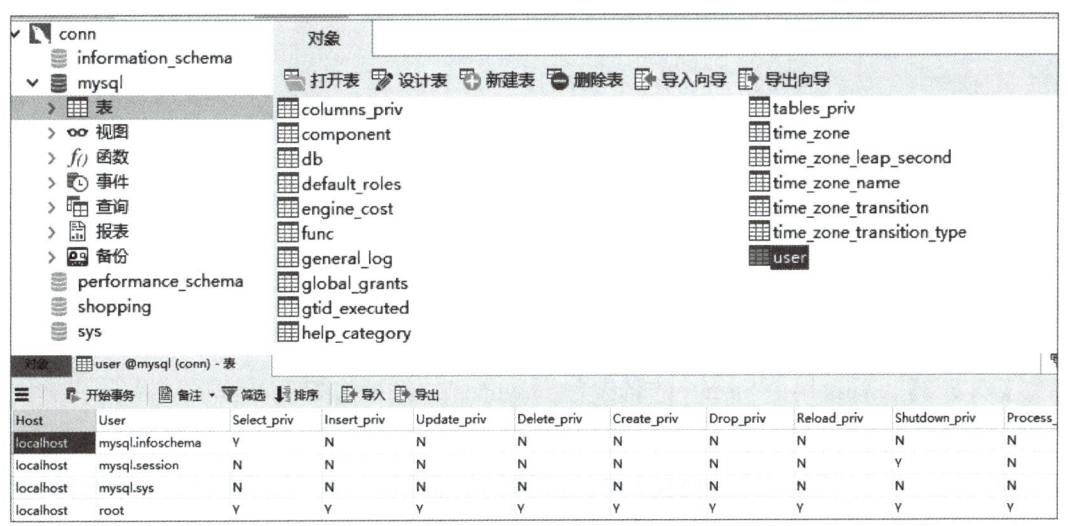

图 6-1　系统数据库自动安装

6.1.2 用户管理

用户管理包括创建用户、修改用户、删除用户等,要实现对用户的管理,就要有相应的操作权限。

1. 创建用户

MySQL 使用 CREATE USER 语句创建用户,并设置相应的密码。其语法格式为:

CREATE USER 账户名称 @ 权限 IDENTIFIED BY 密码

【例 6-1】 创建名称为"user1"的用户,并查看系统中的"user"表。结果如图 6-2 所示。

```
mysql> CREATE USER 'user1';
Query OK, 0 rows affected (0.01 sec)

mysql> SELECT user
    -> FROM mysql.user WHERE USER='user1';
+-------+
| user  |
+-------+
| user1 |
+-------+
1 row in set (0.00 sec)
```

图 6-2 创建名为"user1"的用户

【例 6-2】 创建名为"user2"的用户,设置密码为"123456",且可以在任意地址登录服务器。结果如图 6-3 所示。

```
mysql> CREATE USER 'user2'@'%' IDENTIFIED BY '123456';
Query OK, 0 rows affected (0.00 sec)
```

图 6-3 创建名为"user2"的用户

2. 修改用户

修改用户包括修改用户名称、修改用户密码、修改用户锁定状态和删除用户等。

(1) 修改用户名称

MySQL 使用 RENAME USER 语句修改用户名称,其语法格式为:

RENAME USER 旧账户名 TO 新账户名

【例 6-3】 修改用户"user1"的名称为"newuser"。结果如图 6-4 所示。

```
mysql> RENAME USER 'user1' TO 'newuser';
Query OK, 0 rows affected (0.00 sec)
```

图 6-4 修改用户名称

(2)修改用户密码

在 MySQL 中修改密码的方式有三种：ALTER USER 语句、SET PASSWORD 语句和 mysqladmin 命令。

• 使用 ALTER USER 语句修改用户密码，其语法格式为：

ALTER USER 账户名 @ 权限 IDENTIFIED BY '新密码'

【例 6-4】 修改用户名"user2"的密码为"user123"。结果如图 6-5 所示。

```
mysql> ALTER USER 'user2'@'%' IDENTIFIED BY 'use123';
Query OK, 0 rows affected (0.00 sec)
```

图 6-5 修改用户密码 1

• 使用 SET PASSWORD 语句修改用户密码，其语法格式为：

SET PASSWORD FOR 账户名＝'新密码'

【例 6-5】 修改用户名"user2"的密码为"pwd123"。结果如图 6-6 所示。

```
mysql> SET PASSWORD FOR 'use2'@'%' = 'pwd123';
Query OK, 0 rows affected (0.01 sec)
```

图 6-6 修改用户密码 2

• 使用 mysqladmin 命令修改用户密码，其语法格式为：

mysqladmin -u 用户名 -h 主机地址 -p password 新密码

在使用该命令进行密码修改时，要在"Enter password"提示后输入正确的原密码，才能进行密码修改，一般为了保证连接服务器的安全性，不使用此方式。

(3)修改用户的锁定状态

MySQL 支持 ALTER USER 语句使用 ACCOUNT LOCK 和 ACCOUNT UNLOCK 子句锁定或解锁用户状态。其语法格式为：

ALTER USER 账户名@权限 ACCOUNT LOCK|ACCOUNT UNLOCK

【例 6-6】 锁定用户"user2"。结果如图 6-7 所示。

```
mysql> ALTER USER 'user2'  ACCOUNT LOCK;
Query OK, 0 rows affected (0.00 sec)
```

图 6-7 锁定用户

(4)删除用户

在 MySQL 中，使用 DROP USER 语句删除用户，其语法格式为：

DROP USER 账户名

【例 6-7】 删除用户"user2"。结果如图 6-8 所示。

```
mysql> DROP USER user2;
Query OK, 0 rows affected (0.00 sec)
```

图 6-8 删除用户

6.1.3 权限管理

权限指登录到 MySQL 服务器的用户能对数据库对象执行何种操作的规则集合。在实际应用中,为保障数据的安全,数据库管理员要根据用户的层级进行权限分配。在授予权限或回收权限后,还要通过刷新重新加载权限,否则权限无法生效。

1. 授予权限

在数据库中,不同层级可以授予的权限分为以下四组:

(1)列权限:和表中的一个具体列相关。例如,使用 UPDATE 语句更新表列值的权限。

(2)表权限:和一个具体表中的所有数据相关。例如,使用 SELECT 语句查询表中所有数据的权限。

(3)数据库权限:和一个具体的数据库中所有表相关。例如,在已有的数据库中创建新表的权限。

(4)用户权限:和 MySQL 所有的数据库相关。例如,删除已有的数据库或者创建一个新数据库的权限。

在 MySQL 中可使用 GRANT 语句授予权限,其语法格式为:

GRANT 权限类型 ON 授予权限范围 TO 账户名

【例 6-8】 授予账户"user0"对"shopping"数据库中所有表具有 INSERT、UPDATE、DELETE、SELECT 的权限。结果如图 6-9 所示。

```
mysql> GRANT INSERT,UPDATE,DELETE,SELECT ON shopping.* TO 'user0';
Query OK, 0 rows affected (0.00 sec)
```

图 6-9 授予账户权限 1

【例 6-9】 授予账户"user2"@"localhost"对"shopping"数据库中所有表具有 INSERT、UPDATE、DELETE、SELECT 的权限,并进行查看。结果如图 6-10 所示。

2. 回收权限

回收权限指取消某个用户的特定权限,回收权限可以保障数据库的安全性。在

MySQL 中使用 REVOKE 语句回收用户的权限,其语法格式为:

REVOKE 权限类型 ON 权限范围 FROM 账户名

```
mysql> GRANT SELECT, INSERT, UPDATE, DELETE ON shopping.* TO 'user2'@'localhost';
Query OK, 0 rows affected (0.00 sec)

mysql> SHOW GRANTS FOR 'user2'@'localhost';
+-----------------------------------------------------------------------+
| Grants for user2@localhost                                            |
+-----------------------------------------------------------------------+
| GRANT USAGE ON *.* TO `user2`@`localhost`                             |
| GRANT SELECT, INSERT, UPDATE, DELETE ON `shopping`.* TO `user2`@`localhost` |
+-----------------------------------------------------------------------+
2 rows in set (0.00 sec)
```

图 6-10 授予账户权限 2

【例 6-10】 回收账户"user2"@"localhost"对"shopping"数据库的 SELECT 权限。结果如图 6-11 所示。

```
mysql> REVOKE SELECT ON TABLE shopping.*
    -> FROM 'user2'@'localhost';
Query OK, 0 rows affected (0.00 sec)
```

图 6-11 回收账户权限 1

若要回收用户所有权限,仅需在 REVOKE 语句中加入 ALL PRIVILEGES 关键字即可。其语法格式为:

REVOKE ALL PRIVILEGES,GRANT OPTION FROM 账户名

【例 6-11】 回收账户"user2"的所有权限并使用 FLUSH PRIVILEGES 语句刷新查看。结果如图 6-12 所示。

```
mysql> REVOKE ALL PRIVILEGES,grant option from 'user2'@'localhost';
Query OK, 0 rows affected (0.00 sec)

mysql> FLUSH PRIVILEGES;
Query OK, 0 rows affected (0.00 sec)
```

图 6-12 回收账户权限 2

6.1.4 事务

事务是一组有内在逻辑联系的 SQL 语句,在事务中的操作,要么都执行,要么都不执行,通过事务保证一组数据操作的同步和数据的完整性,使用事务可以提高数据的安全性和执行效率。事务的四个基本原则:原子性、一致性、隔离性和持久性,称为 ACID

原则。

事务的基本操作：开启和提交事务，回滚事务，设置事务保存点。

默认情况下，用户执行的每一条 SQL 语句都会作为单独事务自动提交，若要将一组 SQL 语句当作一个事务操作，则需要执行开启事务、提交事务和回滚事务等操作。在 MySQL 中使用 START TRANSACTION 语句开启事务，其语法格式为：

START TRANSACTION

利用 COMMIT 语句手动提交事务，其语法格式为：

COMMIT

若不想提交事务，可使用 ROLLBACK 回滚事务，其语法格式为：

ROLLBACK

使用 ROLLBACK 语句回滚事务时，事务中所有的操作都会被撤销。

若只需撤销部分操作，可以在事务中设置事务保存点，其语法格式为：

SAVEPOINT 保存点

任务 2 数据的备份与恢复

任务描述

随着信息技术的普及，越来越多的领域使用数据库系统。存储介质损坏、用户误操作、服务器故障、计算机病毒、自然灾害等许多因素会导致数据的丢失或损坏，因此要对数据库系统采取必要措施，保证其发生故障时可以对数据进行恢复，将损失降到最低。本任务主要学习数据库的备份和恢复机制。

6.2.1 数据备份

1. 数据备份概述

数据备份就是对数据库结构、对象和数据建立相应副本，对数据库管理员而言，制定合理的备份策略对于灾难恢复、测试应用、回滚数据修改、查询历史数据、审计等非常有必要。

在实际应用中，可能造成数据丢失的原因有以下五方面：

（1）程序错误。

（2）磁盘故障。

（3）人为操作失误。

（4）自然灾害。

(5) 人为破坏。

根据备份数据的范围来划分,数据备份可以分为完全备份、差异备份和增量备份。

(1) 完全备份:又叫完整备份,是对整个数据库进行备份,保存的是备份完成时刻的数据库,其特点是备份恢复简单,占用磁盘空间大,备份时间较长。

(2) 差异备份:备份第一次完全备份后被修改过的所有文件,其特点是备份数据量大,恢复时仅需要恢复第一次的完全备份和最近一次的差异备份。

(3) 增量备份:以上一次完全备份或者上一次增量备份的时间为节点,只备份时间节点后数据发生变化的部分,其特点是备份数据量小、占用空间小、备份速度快,且恢复时需要上一次完全备份或最近一次增量备份后改变的内容。

2. 使用 Navicat 图形化管理工具备份数据

【例 6-12】 使用 Navicat 备份"shopping"数据库,操作步骤如下:

(1) 打开 Navicat 图形化管理工具并选择"shopping"数据库进行连接,单击"备份"按钮,选择"新建备份"按钮,弹出"新建备份"窗口,如图 6-13 所示。

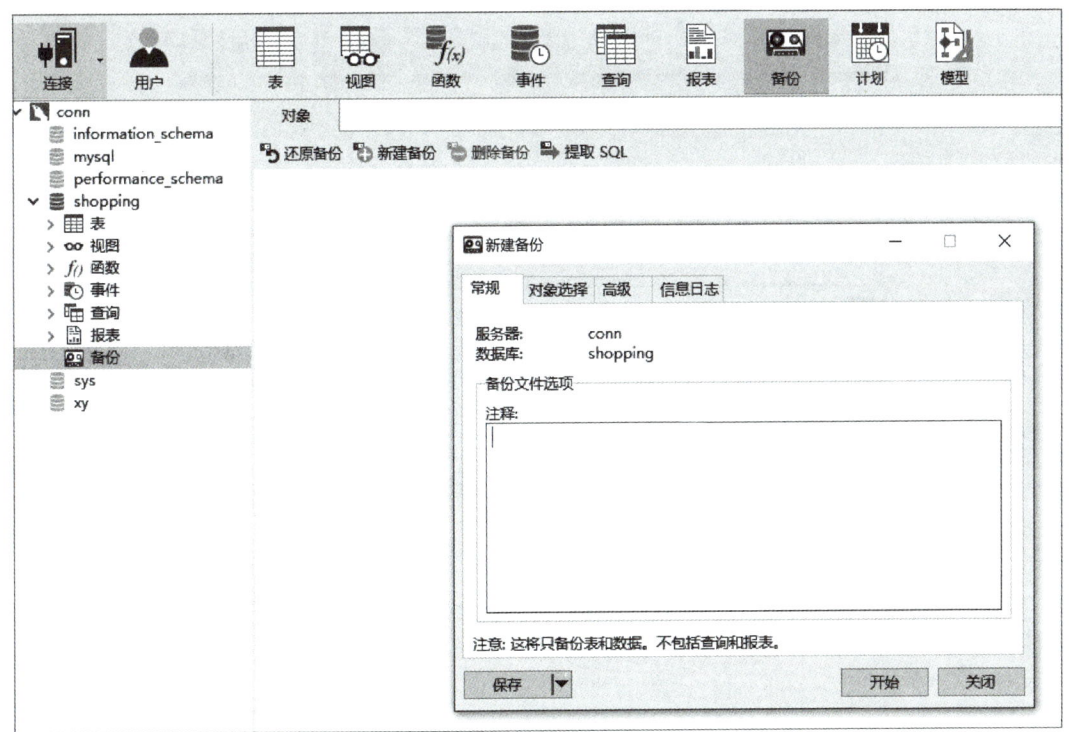

图 6-13 新建备份

(2) 在"对象选择"中勾选需要备份的对象,这里单击"全选"按钮即可,如图 6-14 所示。

(3) 在"高级"选项卡中选择"使用指定文件名"并输入文件名"shopping_backups",如图 6-15 所示。

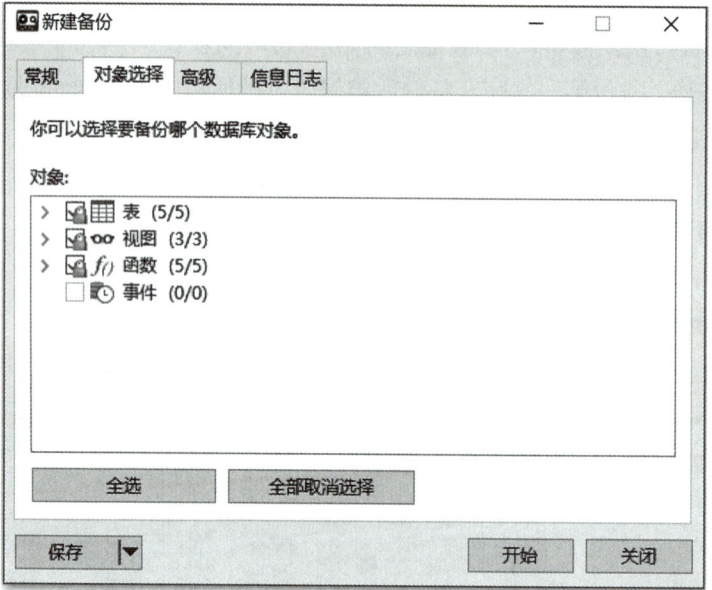

图 6-14 勾选备份对象

图 6-15 输入文件名

(4) 单击"开始"按钮,系统开始执行备份操作,如图 6-16 所示。

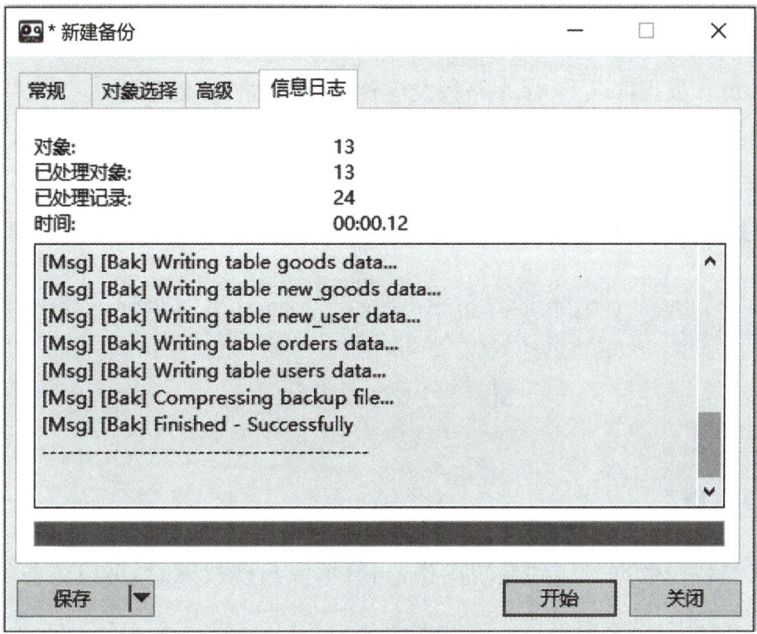

图 6-16　执行备份操作

(5) 备份完毕后,单击"关闭"按钮,在资源管理器中查看备份文件,如图 6-17 所示。

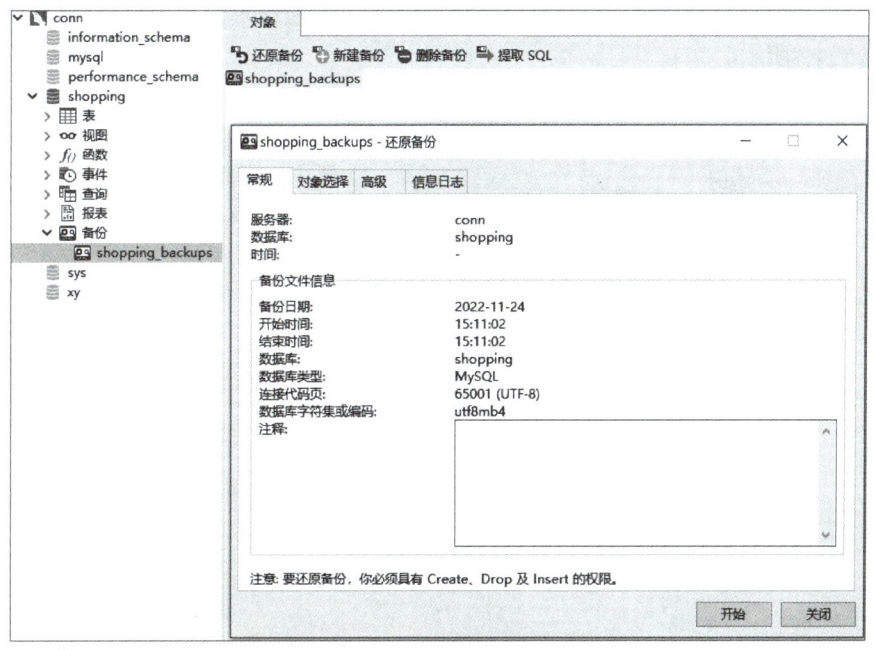

图 6-17　查看备份文件

3. 使用命令备份数据

在 MySQL 中,使用 mysqldump 命令对数据库进行备份分为三种情况:备份一个数据库、备份多个数据库、备份所有数据库。其语法格式分别为:

➢ mysqldump 数据库名称 保存路径及名称 ＞存储路径 文件名

➢ mysqldump －－databases 数据库名 1 数据库名 2……＞存储路径 文件名

➢ mysqldump －－all －databases ＞存储路径 文件名

【例 6-13】 使用本机用户备份"shopping"数据库和"xy"数据库。结果如图 6-18 所示。

```
C:\Users\DELL>mysqldump -u root -p --databases shopping xy >D:\dbs_backups.sql
Enter password: ****
```

图 6-18 备份多个数据库

6.2.2 数据恢复

恢复数据是备份数据对应的系统操作,当数据库出现故障时,可以将备份好的数据库文件加载到服务器中,恢复数据库。

1. 使用 Navicat 图形化管理工具恢复数据

【例 6-14】 使用图形化管理工具 Navicat,将备份文件"shopping_backups"恢复到数据库中,步骤如下。

(1) 打开图形化管理工具 Navicat,并连接到服务器,新建数据库"shopping1",选择"备份",如图 6-19 所示。

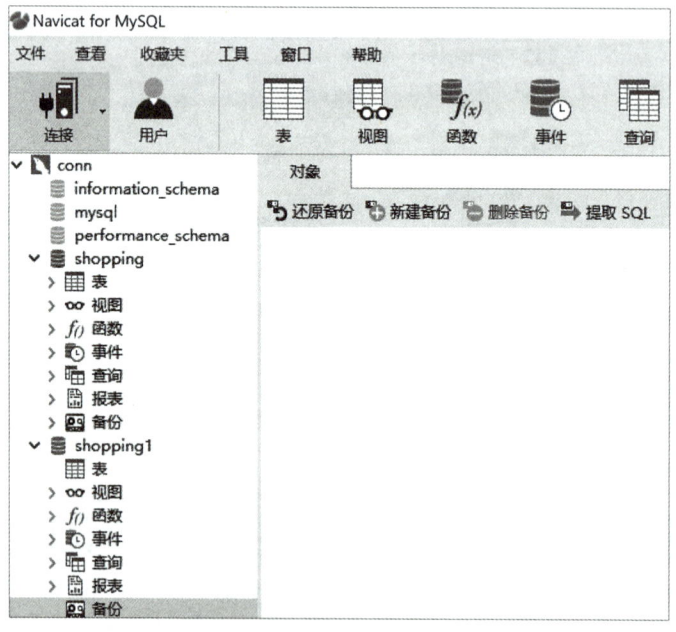

图 6-19 新建数据库

（2）右键单击"备份"选项，选择"还原备份"，输入备份文件名，如图 6-20 所示。

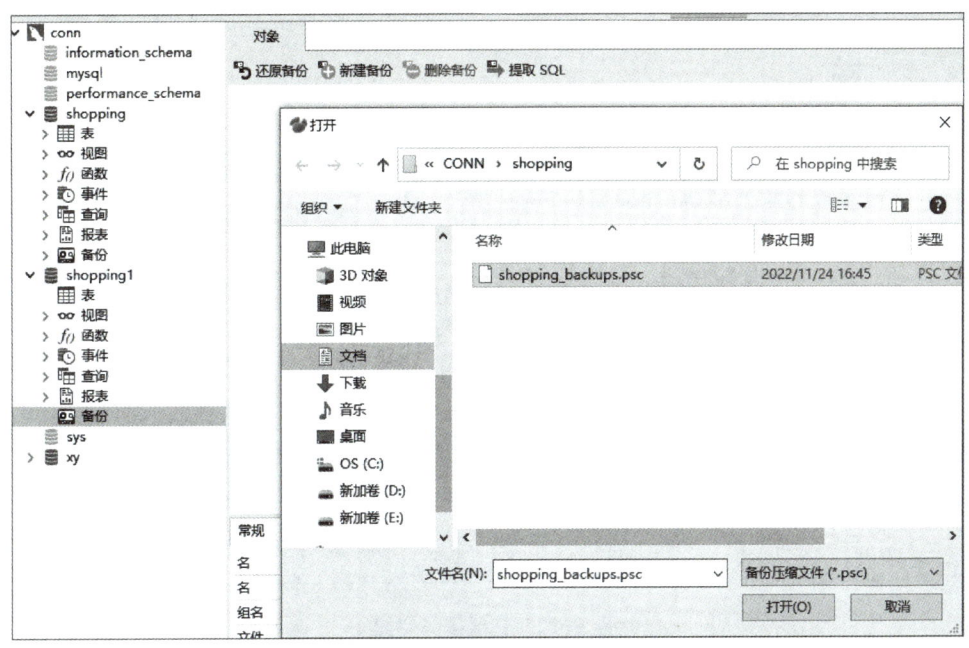

图 6-20　输入备份文件名

（3）打开文件，单击"开始"按钮，执行还原操作，执行完毕后，单击"关闭"按钮，完成数据恢复，如图 6-21、图 6-22 所示。

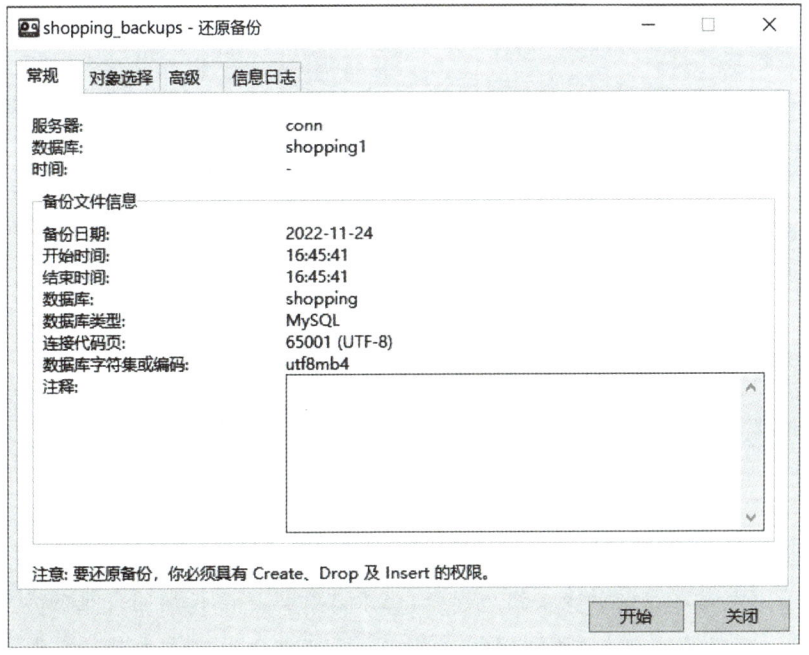

图 6-21　执行还原操作

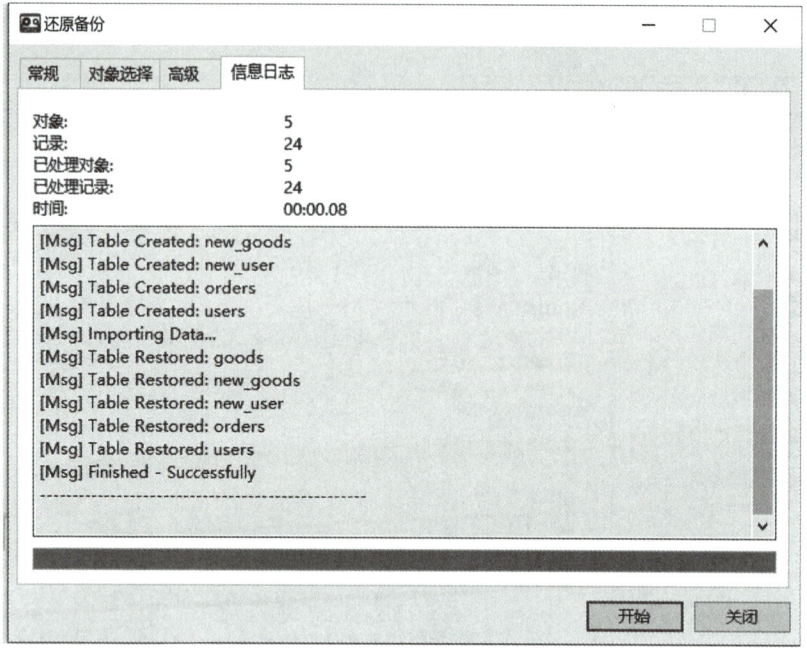

图 6-22　还原操作完成

2. 使用命令恢复数据

在 MySQL 中可以使用命令恢复数据,其语法格式为:

mysql -u 用户名 -p 密码 ＜ 存储路径 文件名

【例 6-15】　使用命令将备份文件"shopping_backups"恢复到数据库"xy"中。结果如图 6-23 所示。

```
C:\Users\DELL>mysql -u root -p xy <D:\dbs_backups.sql
Enter password: ****
```

图 6-23　恢复数据

任务 3　使用日志备份和恢复数据

任务描述

日志是数据库管理系统的重要组成部分,记录着数据库运行期间发生的各种事件。当数据库出错时,可以通过日志文件查看出错原因,还可以通过日志文件进行数据恢复。本任务主要学习各种日志的作用和使用方法,使读者能够利用日志文件对数据库进行维护。

6.3.1 日志概述

日志是将数据库中的每一个操作与变化产生的信息记录到专用文件中,又称为日志文件。在日志中可以查询到数据库的运行、用户操作、错误信息等,日志文件通常存储在数据目录下,为数据库的管理和优化提供必要信息。

MySQL 中的日志主要包括错误日志、二进制日志、通用查询日志和慢查询日志四种。使用日志可以提高系统安全性,但启动日志会降低数据库的性能,尤其是启动查询日志,数据库服务器会花费较多时间进行日志信息记录,而且日志文件也占用较大存储空间,因此,默认情况下,MySQL 服务器仅启动错误日志。

6.3.2 错误日志

错误日志主要记录 MySQL 服务的启动、运行或停止时出现的问题,当数据库发生故障、无法正常使用时,可以首先查看此日志。在 MySQL 中,错误日志默认是开启的,并且该日志类型无法被禁止。

1. 查看错误日志

MySQL 中的错误日志是以文本文件形式存储的,因此可以使用文本编辑器直接查看,也可以使用 SHOW VARIABLES 语句先查看错误日志的存储路径。

【例 6-16】 查询错误日志文件的存储路径。结果如图 6-24 所示。

```
mysql> SHOW VARIABLES LIKE 'log_error';
+---------------+------------------------+
| Variable_name | Value                  |
+---------------+------------------------+
| log_error     | .\DESKTOP-RNG1160.err  |
+---------------+------------------------+
1 row in set, 1 warning (0.00 sec)
```

图 6-24 查看错误日志

2. 删除错误日志

错误日志本身是以文本文件形式存储的,因此可以直接删除,但在运行状态下删除日志文件后,MySQL 并不会自动创建新的日志文件。可以通过 FLUSH LOGS 语句重新加载。结果如图 6-25 所示。

```
mysql> FLUSH LOGS;
Query OK, 0 rows affected (0.06 sec)
```

图 6-25 重新加载日志

6.3.3 二进制日志

二进制日志记录了所有数据定义语句和数据操作语句对数据库的更改操作,语句以事件的形式存储,并记录了语句发生的时间、执行时长、操作数据等,因此二进制日志是基于时间点进行恢复的,对数据损坏后的恢复至关重要。

1. 启动和设置二进制日志

默认状态下,二进制日志是关闭的,可用修改 MySQL 的配置文件 my.ini 的方法来设置二进制日志,设置完成后重新启动 MySQL 服务,二进制日志信息才可生效。可先在 my.ini 中查找相关参数的位置,如图 6-26 所示。

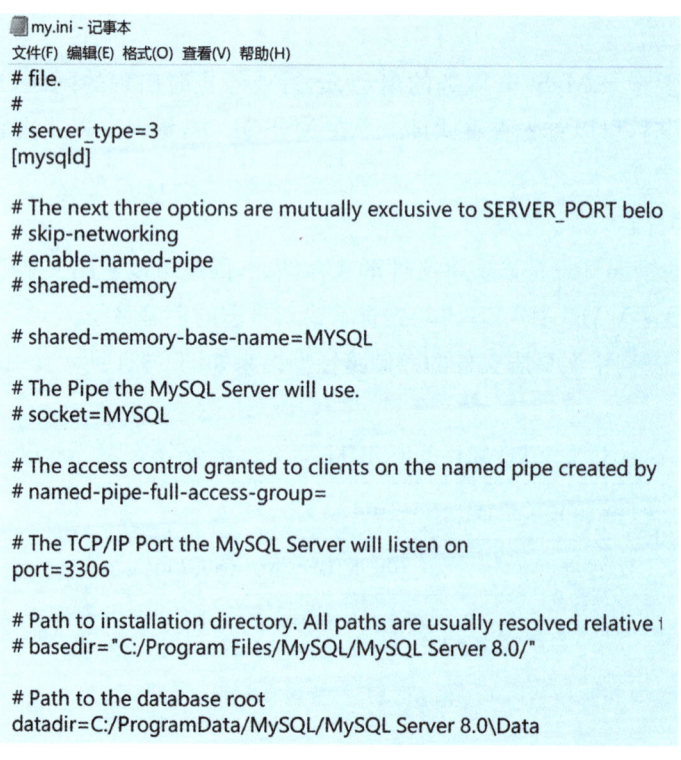

图 6-26 查找参数位置

启动二进制日志的操作步骤如下:
(1) 在配置文件中添加相关语句并保存,如图 6-27 所示。

图 6-27 添加语句

（2）重新启动 MySQL 服务，系统自动生成文件，且每次重新启动都会生成新的二进制日志文件，并存储在 MySQL 数据目录下。

（3）用 SHOW VARIABLES 语句查看日志设置情况，如图 6-28 所示。

```
mysql> SHOW VARIABLES Like 'log%';
+------------------------------------------+-----------------------------------------------------+
| Variable_name                            | Value                                               |
+------------------------------------------+-----------------------------------------------------+
| log_bin                                  | ON                                                  |
| log_bin_basename                         | C:\ProgramData\MySQL\MySQL Server 8.0\Data\logbin   |
| log_bin_index                            | C:\ProgramData\MySQL\MySQL Server 8.0\Data\logbin.index |
| log_bin_trust_function_creators          | OFF                                                 |
| log_bin_use_v1_row_events                | OFF                                                 |
| log_error                                | .\SHJ.err                                           |
| log_error_services                       | log_filter_internal; log_sink_internal              |
| log_error_suppression_list               |                                                     |
| log_error_verbosity                      | 2                                                   |
| log_output                               | FILE                                                |
| log_queries_not_using_indexes            | OFF                                                 |
| log_raw                                  | OFF                                                 |
| log_replica_updates                      | ON                                                  |
| log_slave_updates                        | ON                                                  |
| log_slow_admin_statements                | OFF                                                 |
| log_slow_extra                           | OFF                                                 |
| log_slow_replica_statements              | OFF                                                 |
| log_slow_slave_statements                | OFF                                                 |
| log_statements_unsafe_for_binlog         | ON                                                  |
| log_throttle_queries_not_using_indexes   | 0                                                   |
| log_timestamps                           | UTC                                                 |
+------------------------------------------+-----------------------------------------------------+
21 rows in set, 1 warning (0.00 sec)

mysql>
```

图 6-28　查看日志设置情况

2. 查看二进制日志

在 MySQL 中，可以通过 SHOW BINARY LOGS 语句查看二进制日志个数和文件名，其中，日志个数与启动次数相同，每启动一次 MySQL 服务就会产生一个新的日志文件。

【例 6-17】使用 SHOW BINARY LOGS 语句查看当前二进制日志的文件信息。结果如图 6-29 所示。

```
mysql> SHOW BINARY LOGS;
+----------------------------+-----------+-----------+
| Log_name                   | File_size | Encrypted |
+----------------------------+-----------+-----------+
| DESKTOP-RNG1160-bin.000001 |       180 | No        |
| DESKTOP-RNG1160-bin.000002 |      1334 | No        |
| DESKTOP-RNG1160-bin.000003 |     23209 | No        |
| DESKTOP-RNG1160-bin.000004 |       180 | No        |
| DESKTOP-RNG1160-bin.000005 |     12651 | No        |
| DESKTOP-RNG1160-bin.000006 |       180 | No        |
| DESKTOP-RNG1160-bin.000007 |     15659 | No        |
| DESKTOP-RNG1160-bin.000008 |     18003 | No        |
| DESKTOP-RNG1160-bin.000009 |       180 | No        |
| DESKTOP-RNG1160-bin.000010 |       157 | No        |
+----------------------------+-----------+-----------+
10 rows in set (0.00 sec)
```

图 6-29　查看当前二进制日志的文件信息

另外,还可以使用mysqlbinlog命令查看二进制日志文件的内容(图6-30),其语法格式为:

mysqlbinlog(可选参数)二进制日志文件

```
C:\Users\DELL>mysqlbinlog -v logbin.000001
# The proper term is pseudo_replica_mode, but we use this compatibility alias
# to make the statement usable on server versions 8.0.24 and older.
/*!50530 SET @@SESSION.PSEUDO_SLAVE_MODE=1*/;
/*!50003 SET @OLD_COMPLETION_TYPE=@@COMPLETION_TYPE,COMPLETION_TYPE=0*/;
DELIMITER /*!*/;
mysqlbinlog: File 'logbin.000001' not found (OS errno 2 - No such file or directory)
ERROR: Could not open log file
SET @@SESSION.GTID_NEXT= 'AUTOMATIC' /* added by mysqlbinlog */ /*!*/;
DELIMITER ;
# End of log file
/*!50003 SET COMPLETION_TYPE=@OLD_COMPLETION_TYPE*/;
/*!50530 SET @@SESSION.PSEUDO_SLAVE_MODE=0*/;
```

图6-30 使用mysqlbinlog命令查看二进制日志文件的内容

3. 删除二进制日志

二进制日志文件能记录用户对数据的修改操作,但随着文件的不断增长,其会影响数据库的性能,因此过期的二进制日志要及时删除。MySQL提供了两种删除二进制日志的语句。

(1)使用RESET MASTER语句删除所有二进制文件。结果如图6-31所示。

```
mysql> RESET MASTER;
Query OK, 0 rows affected (0.05 sec)
```

图6-31 删除所有二进制文件

(2)使用PUREG MASTER/BINARY LOGS语句删除指定的二进制日志文件。PUREG MASTER指删除文件名编号比指定文件名编号小的所有二进制日志文件;BINARY LOGS指删除指定时间点之前的所有二进制日志文件,其语法格式为:

PURGE MASTER/BINARY LOGS TO 二进制日志文件名

6.3.4 通用查询日志

不管查询或命令是否有结果,通用查询日志都会记录服务器接收到的每一个查询或命令,因此,启动通用查询日志会产生很大的系统开销。在默认情况下,MySQL服务器不开启通用查询日志,只有需要跟踪某些特殊SQL性能问题时才会将其启动。通用查询日志文件以".log"为后缀,若在配置文件my.ini中没有指定文件名,默认文件名为主机名。通用查询日志的主要功能并不是为了恢复数据,而是为了监控用户的操作。

1. 启动和设置通用查询日志

在MySQL中,可以通过修改系统配置文件my.ini来启动通用查询日志,其配置信息设置如图 6-32 所示。

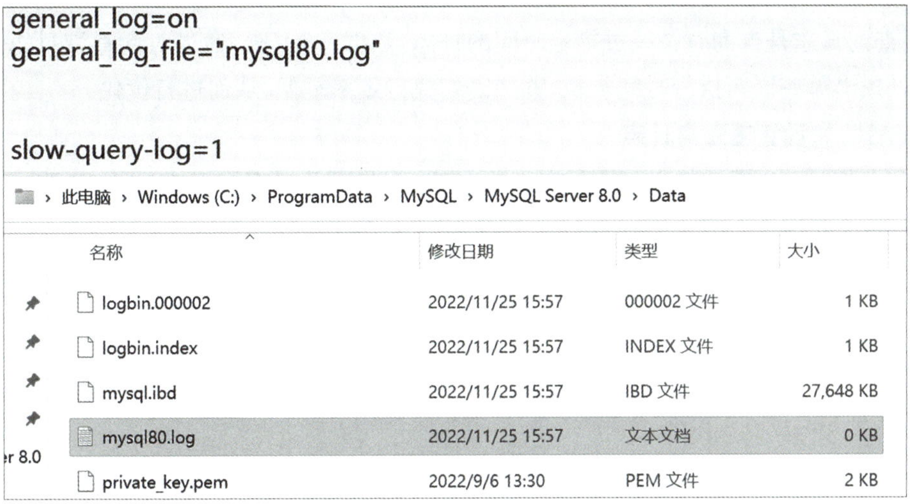

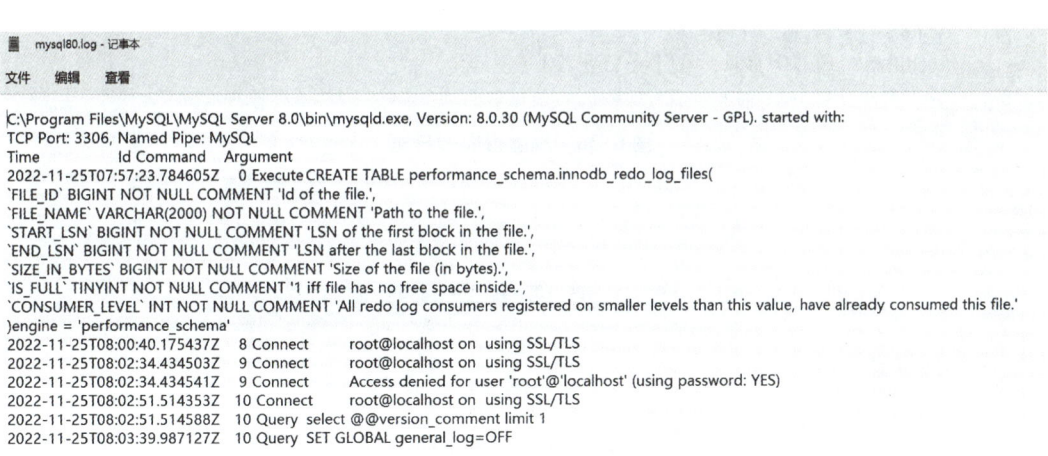

图 6-32　启动通用查询日志

2. 查看和关闭通用查询日志

在 MySQL 中，可通过 SHOW VARIABLES 语句或直接打开 MySQL 目录下的 mysql80. log 文件查看通用查询日志，如图 6-33 所示。

图 6-33　查看通用查询日志

在 MySQL 中使用 SET 语句可关闭查询日志。结果如图 6-34 所示。

```
mysql> SET GLOBAL general_log=OFF;
Query OK, 0 rows affected (0.02 sec)
```

图 6-34　关闭查询日志

6.3.5 慢查询日志

慢查询日志是查询时长超过指定时间的日志,数据库管理员通过对慢查询日志进行分析,可以找出哪些语句执行时间较长,哪些语句执行效率较低,从而进行优化。

1. 启动和设置慢查询日志

在 MySQL 中,默认的慢查询日志是关闭的,若要启动慢查询日志,可以修改配置文件 my.ini,其配置信息如图 6-35 所示。

```
#long_query_time=10
Slow-query-log=ON
slow_query_log_file=mysql80-slow.log
long_query_time=5
log-queries-ot-using-indexes=ON
```

图 6-35　启动慢查询日志

2. 查看和删除慢查询日志

慢查询日志以文本文件形式存储,因此可以使用文本编辑器查看。

慢查询日志可以直接删除。删除后在不重启 MySQL 服务器的情况下,在客户端执行 FLUSH LOGS 语句即可重建日志文件,如图 6-36 所示。

```
mysql> FLUSH LOGS;
Query OK, 0 rows affected (0.06 sec)
```

图 6-36　删除慢查询日志

课后习题

一、选择题

1. 以下哪个语句用于撤销权限?(　　)

　　A. DELETE　　B. DROP　　C. UPDATE　　D. REVOKE

2. 创建用户的语句是(　　)。

　　A. CREATE USER　　　　　　B. INSERT USER

　　C. CREATE root　　　　　　D. MYSQL user

3. MySQL 中,使用(　　)语句来为指定的数据库添加用户。

　　A. CREATE USER　B. UPDATE　　C. GRANT　　D. INSERT

4. 备份 MySQL 数据库的命令是(　　)。

　　A. mysqldump　　B. copy　　C. backup　　D. mysql

5. 在某一次完全备份的基础上,只备份其后数据变化的备份类型称为(　　)。

　　A. 比较备份　　　B. 差异备份　　　C. 完全备份　　　D. 增量备份

6. 在 MySQL 内部有四种常见的日志,(　　)不能直接使用文本编辑器查看日志内容。

　　A. 二级制日志　　B. 错误日志　　　C. 通用查询日志　　D. 满查询日志

7. 查看和恢复二进制日志的命令是(　　)。

　　A. mysqldump　　B. mysqlbinlog　　C. mysqllimport　　D. mysql

二、简答题

1. 如何使用日志文件进行数据备份?
2. 简述 MySQL 数据库中四种日志文件的优缺点。

项 目 实 训

1. 实训任务

(1) 创建用户。

(2) 授予用户权限。

(3) 备份、恢复数据库。

(4) 使用四种日志文件。

2. 实训目的

(1) 能使用 SQL 语句创建用户。

(2) 能使用 SQL 语句设置用户权限。

(3) 能使用 SQL 语句修改用户权限。

(4) 能分别使用 Navicat 和命令行备份和恢复数据库。

(5) 能设置、查看、删除四种日志文件。

3. 实训内容

(1) 使用 SQL 语句创建一个用户"zhouyang",密码为"123456"。

(2) 使用 SQL 语句创建一个用户"zhanglin",密码为"123456",授予该用户对"shopping"数据库的 SELECT 权限。

(3) 使用 SQL 语句回收"zhanglin"对"shopping"数据库的 SELECT 权限。

(4) 分别使用 Navicat 和 mysqldump 命令备份"shopping"数据库。

(5) 分别使用 Navicat 和 mysql 命令恢复"shopping"数据库。

(6) 设置并启动二进制日志,指定文件名为"logbin.01"。

(7) 使用 mysqlbinlog 命令查看二进制日志文件。

参考答案

项目一　MySQL 数据库基础

一、选择题

1. C　2. B　3. B　4. C　5. D　6. A

二、简答题

1. 常见的关系型数据库管理系统有 MySQL 数据库管理系统、Oracle 数据库管理系统、DB2 数据库管理系统和 SQL Server 数据库管理系统等。

（1）MySQL

MySQL 是开放源码的数据库管理系统，MySQL 具有性能高、成本低、可靠性好、可跨平台等特点，目前广泛应用于互联网行业。

（2）Oracle

Oracle 数据库系统是目前世界上最流行的关系型数据库，不仅具有完整的数据管理功能，还是一个分布式数据库系统，支持各种分布式功能。Oracle 具有可移植性好、使用方便、开发工具的界面友好等特点，功能齐全。

（3）DB2

DB2 是国际商业机器（IBM）公司出品的关系型数据库管理系统，具有较好的可伸缩性。从大型机到单用户环境均可支持，应用于常见的服务器操作系统平台。DB2 具有很好的网络支持功能，适用于大型分布式应用系统。

（4）SQL Server

SQL Server 是微软（Microsoft）公司推出的关系型数据库管理系统，具有可伸缩性、可用性、可靠性等特点，使系统管理和数据库管理更加直观、简单。

2. 数据库是按照一定存储结构相互联系的数据集合，用于支持应用系统运行所必要的数据。数据库管理系统是操纵和管理应用系统数据库的软件，可以实现对数据库中对象和数据的添加、修改和删除，保证数据库的安全性和完整性。

数据库系统是由以数据库管理系统为核心的软件、数据库和数据库管理员（DBA）组成的系统，它包含数据库及数据库管理系统。

项目二 操作数据库与数据表

一、选择题
1. A 2. C 3. D 4. B 5. B 6. D 7. B 8. D

二、简答题
1. 使用 Datetime。因为在图书管理系统中，图书借阅时间与还书时间和超期罚款计算有关，所以借阅时间要精确。

2. NULL 与 0 不同。在 MySQL 中，NULL 与任何值不相等，NULL 代表不确定；而"0"代表值为 0。

项目三 查询系统数据

一、选择题
1. B 2. D 3. D 4. C 5. A 6. D 7. A 8. D

二、简答题
1. 二者在语法上表现不相同，WHERE 子句在语法上放在 FROM 子句后，用来筛选表中满足条件的数据行；HAVING 子句需与 GROUP BY 子句一起使用，用来筛选分组统计后的数据行，通常用来筛选聚合数据的行。

2. 连接查询和子查询是多表查询的不同方式。连接查询是使多张表进行笛卡尔集后再对关联列进行等值筛选，该种查询结构上较简单，方便易用。子查询则可以根据查询逻辑将内层查询的结果作为外层查询的条件、表达式和相关性查询，子查询更灵活，它还可以嵌套在 INSERT、UPDATE、DELETE 语句中。一般情况下，使用连接查询的 SELECT 语句都可以用子查询来实现。子查询实现的查询业务也可以写成连接查询。

项目四 优化系统数据

一、选择题
1. D 2. B 3. D 4. D 5. B 6. C

二、简答题
1. 数据量小的表不要使用索引，因为通过索引查询记录可能比直接扫描整张表还要慢；不要建立过多的索引，索引并非越多越好，过多的索引会占用大量磁盘空间，并且会降低操作的性能；对于经常执行修改操作的表不要创建过多索引，且索引中的列也应尽可能少，而对于经常执行查询操作的字段，应该创建索引；在条件表达式中经常会用到的不同值较多的列上创建索引，不同值较少的列不要创建索引；在频繁进行排序或分组的列上创建索引。

2. 视图保障数据安全性：为不同的用户定义不同的视图，可以限制用户的访问范围。通过视图机制把需要保密的数据对无权存取这些数据的用户隐藏起来，可以对数据库提供一定程度的安全保护。

项目五 数据库编程

一、选择题

1. A 2. B 3. D 4. C 5. A

二、简答题

1. 优点：(1)存储过程执行一次后，其执行计划就保留在缓存中，再次调用只需要从高速缓冲存储器中调用即可，提高了系统的性能；(2)存储过程的数据返回可以通过 SELECT 语句和输出函数实现；(3)存储过程的应用灵活，且可以嵌套在触发器或事件中；(4)数据库的管理员可以对存储过程进行单独的权限控制，避免非授权用户对数据的访问，从而保证数据的安全性。

缺点：(1)可移植性差；(2)开发调试复杂，由于电子集成驱动器(IDE)的问题，存储过程的开发调试比一般程序更困难；(3)SQL 本身是结构化查询语言，不是面向对象语言，在复杂业务的处理上会比较吃力。

2. (1)MySQL 触发器始终是基于表中的一条记录触发，而不是基于一组 SQL 语句。因此，如果需要变动整个数据集而数据集的数据量又较大时，触发器效率会非常低；(2)触发器是由于表中的数据变化引起另外的数据表中发生相应的数据变化，过多的触发器会导致数据的链式反应，效率很低，甚至造成死循环，这些都是使用触发器需要注意的。

项目六 数据库系统的安全性和可用性

一、选择题

1. D 2. A 3. A 4. A 5. B 6. A 7. B

二、简答题

1. (1)启动和设置二进制日志：在 my.ini 配置文件的[mysqld]组下添加如下语句：log-bin="logbin"；(2)重启 MySQL 服务；(3)执行 SHOW VARIABLES 语句，查看日志设置情况；(4)查看二进制日志文件的存储路径；(5)使用 mysqlbinlog 命令查看二进制日志文件的内容。

2. (1)错误日志：能够实现数据库排错；记录启动、运行或停止时出现的问题，一般也会记录警告信息。

(2)查询日志：能够实现数据库调试；记载着 MySQL 的所有用户操作，包括启动和关闭服务、执行查询和更新语句等。

（3）慢查询日志：能够记录执行时间超过指定时间的查询语句。记录所有执行时间超过 long_query_time 的所有查询或不使用索引的查询，可以定位服务器性能问题。

（4）二进制日志：以二进制文件的形式记录数据库中所有更改数据的语句，以及任何引起或可能引起数据库变化的操作，主要用于复制和即时点恢复，且不依赖于存储引擎类型。

参考文献

[1] 李锡辉,王敏.MySQL数据库技术与项目应用教程[M].北京:人民邮电出版社,2022.
[2] 马洁,郭义,罗桂琼.MySQL数据库应用案例教程[M].北京:航空工业出版社,2018.
[3] 周德伟,覃国蓉.MySQL数据库基础实例教程(微课版)[M].北京:人民邮电出版社,2017.
[4] 郑小蓉,段萍.MySQL数据库项目化教程[M].北京:中国水利水电出版社,2019.